Str
HORROR STORIES

- Street rod wrecks and crashes!
- '51 Chevy parks on a '55!
- Race car hauler blows a tire!
- Fishtailing disaster!
- Cars flooded at the Nationals!
- '31 roadster meets a stone wall!
- A skunk stinks up a cruise!

And many more!

Also...many hints on how to avoid such calamities!

Thank you Mike and Suzanne! Hope you enjoy especially the introduction about my early interest in hot rod cars...

Curt Stultz

All rights reserved. No part of this publication may be reproduced, in any form or by any means, without the permission in writing from the publisher.

Street Rod Horror Stories
ISBN 978-0-578-47470-0
Printed in the United States of America

Village Copy & Print
1 Town Market Place, Ste. 12
Essex Junction, Vermont 05452
info@villagecopy.com
802-879-4426

Street Rod Horror Stories

is dedicated to my sister

Mariel Stultz Klingbeil

Mariel Klingbeil creates and exhibits her art at her home studio in Callaway, Nebraska.

Mariel has had a passion for drawing, painting and sculpting hundreds of works of art since she was very young. I have always admired her great desire to create.

She truly followed her dream. She now lives in Nebraska and is a well-known Western artist. She was inducted into the Sand Hills Cowboy Hall of Fame in 2016. Her work features her favorite subject: the western horse. She also paints ranch and cowboy scenes as well as portraits and makes extraordinary bronze sculptures.

She draws pretty good cartoons of street rods as well!

You will find many of her cartoons within the pages of Street Rod Horror Stories.

Contents

Introduction The Making of a Street Rodder	Art Stultz	1
Bridging the Gaps	Bob White	19
Getting the Shaft	Don Amundson	21
Insulin Insult	John Cordischi	22
A Six Day Trip	Art Stultz	23
Clearly the Wrong Tubing	Ken Bessette	25
Gasoline and Droplights Don't Mix!	Paul Benoit	26
Off to a Bad Start	John Coolidge	28
18-Wheeler to the Rescue	Dennis O'Brien	29
A Stinky Situation	Bill Lyon	29
Thank You Fellow Pages…Twice!	Larry Anderson	30
Snowbanks Win Every Time	Kevin O'Brien	32
First Street Rod	Art Stultz	32
Battery and Assault	Kevin O'Brien	36
Jersey Barriers and Pennsylvania Roads	Anonymous	37
Flickers and Flames	Al Downing	38
French Hill Fish Tale	Dennis Caron	39
A Tale with a "Nashty" Ending	Allen Haines	41
Do What Ya Gotta Do	Roger Dussault	43
The Angry Trailer	Walt Fuller	45
Reaching High	Bob Brunette	46
A Dip in the Pool	Mark Gissendaner	47
Difficult Decisions	Allen Haines	48
Déjà vu All Over Again	Mario Fortin	49
Holy Piston	Dennis O'Brien	50
Pumps in a Barrel	Art Stultz	51
Minor Details	Frank Hanley	53
The Obstacle Course	Don Amundson	55
Street Rod Farmer	Norm Leduc	56
A Busted Generator	Doug Hughes	57
Bikes Have Floorboards Too!	Brian Martin	58
Wiped Out on the Big Trip	Jim Higgins	60
Oh Deer!	Stan Morrow	60
Street Rod Wedding	Mike Burnham	62
Rumble on the Highway	Stephanie Paris	63
Oh Canada! Trouble x 8	Art Stultz	64

Contents

Dad Was Mad - Story #1 Racing Rudy	Ken Bessette	68
Dad Was Mad - Story #2 Over the Wall	Ken Bessette	69
Dad Was Mad - Story #3 Rocket Man	Ken Bessette	70
Had to Bag This Trip	Bob Powell	71
"Don't Be Judging People"	Jim Rowlett	72
Real Flames	Walt Kruger	73
Tragic Fraying	Gene Tinney	75
Bearing Down on a Problem	Dan Tourigny	76
The Great Flood of 1992	Roger Dussault	77
Gas Pains	Bob Brunette	79
Shimmy, Shimmy, Shake Shake Shake	George Tebbetts	80
Gimme a Brake, not a Break	Glenn Turner	81
An Ounce of Prevention	Don Amundson	82
A Trip to the Market	Art Stultz	85
A Bent Cam Pin	E and J Wrightington	86
Leaves in the Dumpster	Dennis O'Brien	87
Old Time Street Rodder	A. Stultz-R. Kendrew	88
My Shoes are Showing!	Dennis O'Brien	91
Not Dead on the Road Yet	Fred Hout	92
Hot Shot Studebaker	Ken Bessette	93
Saving Buddy	Art Stultz	95
The Grasshopper	John Smith	97
Lobsters Paid the Bill	Ed Miller	98
I Shoulda Gone to Church	Stan Morrow	99
A Hauler Horror	Charlie Bryson	101
New is Not Always New	Jed Greeke	103
Tired of this Trip	Bill Lyon	105
A Great Save at The Great Race	Art Stultz	107
Nuts to You	Lyle Smith	108
Shuffle Off to Buffalo; Limp Back Home	Art Stultz	109
Pumpin' Right Along...Maybe!	Jim Knack	111
Takin' it for Granted	Bob Powell	112
Yankee Ingenuity...in Pennsylvania	Jim Ricker	113
Air Conditioning...at Any Cost!	Walt Fuller	115
Island Hopping	Ken Bessette	116
Bearings, Crashes and Burning Oil	Art Stultz	117
Whoa Effie!	Dan Sargent	119

Contents

The Invincible Wall	Roger Kendrew	121
The Price of a Free '55 Chevy	George Lucia	123
Encounter with a Stone Wall	D. Juonis-D. Bishop	125
Crunching the Best	Jim Ricker	126
A Damper on My Trip	Paul Zampieri	128
Loose Nuts: Preach What I Teach	Art Stultz	129
Slippin' and a Slidin'	Lionel "Puddy" Paris	131
Triple Axle Debacle	Wings Kalahan	132

Introduction

The Making of a Street Rodder

By Art Stultz

Many people the world over have a passion for automobiles. In the United States there is an innate love for the look and performance of a sleek smooth Camaro, Mustang or Dodge Viper. In Europe it might be a Porsche Spyder or a Lamborghini that catches people's fancy. Getting behind the wheel of a high-performance car excites the driver with the feeling of power and control. He senses the potential of controlling a huge assemblage of metal, rubber and glass with his hands, feet, and mind.

The early pioneers of the automobile, such as Benz, Daimler, Levassor, Otto, Ford and others, no doubt were not thinking so much of speed and power but of only simple transportation. Wouldn't it be a great invention to build a contrivance that would take the place of the horse and carriage with its many limitations?

With the huge success of dozens of early pioneers, the automobile became a reality and people became enamored with the performance and increasing reliability of the automobile at the dawn of the 20th century. The quest for speed and power was a natural extension of those who designed, built and enjoyed driving them.

My Introduction to Automobiles

I was lucky to get in on the early days of the development of the automobile. I was born in 1936 so the automobile was just beyond its infancy and into the age where the world depended on the automobile and the truck for daily commuting and transportation of goods. The horse-and-carriage was relegated to a secondary position in the lives of people.

Most of us try to determine what was the earliest thing they can remember in their lives. For me, it involved a blue metal toy car in the dirt driveway of my family's rented home in Burlington, Vermont. I am not sure what year it was, but knowing that we lived there only a year or so it must have been about 1939 that I was able to play with this basic little toy that now that I know makes and models of cars-resembled a 1939 Chevrolet four-door sedan. I pushed it up and down the driveway. Probably got the car and myself pretty dirty.

Step One: A Tricycle

We were in our permanent home in 1943 when I got a tricycle. I was proud that my dad bought me a "Super Model," let's call it, rather than the traditional kids' tricycle with the pedal and bell crank as part of the front axle. Mine had the pedals between the front wheel and rear wheels with the pedals driving a sprocket that in turn drove the rear wheels by chain drive. Cool! I drove that tricycle for all it was worth up and down the sidewalks where I lived.

One springtime the snow melted, leaving the sidewalks clear, but there were various snow banks remaining that I could drive my 'cycle into. What fun! My sister and I had built a huge snowman that winter and his remains were still there, having been hardened with thawing and freezing of the March nights and days. I would run the front wheel headlong into the snowman trying to break him apart with little luck. Okay... get a faster approach and hit it harder! I was quickly becoming a hot rodder with my faster and stronger collisions into the big ball of snow. In due time I discovered that the front fender was striking the frame as I made left and right turns. That was strange. As a seven-year-old I should have known that I had bent the fork backward but I didn't. I asked Dad why this had happened. He figured it out in short order and chastised me for treating my trike so badly. He helped me straighten out my front fork and get my ad hoc hot rod back on the sidewalk.

Meanwhile World War II had been going on for several years. Metal scrap and newspapers were collected all over the country to be converted for use in the war effort. My dad's big beautiful '39 Buick that I loved was "put up on blocks" since gasoline was rationed making it difficult to do much driving. He rode a bicycle on which he had mounted a peach basket to carry groceries. Hot rodding was just about impossible in those days what with the country so involved with simply surviving. The country's hot rods were in the form of jeeps, tanks, halftracks and a myriad of other configurations of war-effort vehicles.

Undoubtedly many a soldier serving in World War II thought of the Army Jeep

as a little hot rod that could scoot adeptly over hill and dale, through the woods and across the beaches. Many a jeep perhaps whetted the appetites of the American GI's who longed to own a hot rod in peacetime.

The war was finally over in 1945 and we were able to buy new cars and were free to buy gasoline for them without the use of government rations. I went for a ride with my neighbor in his '40 Chevrolet and he handed the gas station attendant actual money, not ration tickets. How novel!

I wonder how many hot rodders and street rodders have really thought about the fact that we do not have any 1943, '44 or '45 model cars around. The war effort went solely toward military vehicles.

A World War II Jeep CJ-2A

A Pleasant Surprise: A Bicycle

A full-sized bicycle was the next natural progression for a mechanically minded thirteen-year-old in 1949. I was surprised to come home from school one day and here was a used bicycle parked in the garage. I just knew it was mine before Dad announced that he found one in town at a good price. I was off and running with the next era of my hot-rodding life. I dolled up my bike (manufacturer unknown) with all kinds of decorations such as mud flaps, reflectors, horns and lights. My uncle was a truck driver so he gave me some discarded truck "clearance lights" that I used as tail lights. I mounted a heavy six-volt lantern battery under the seat. A friend tried to tell me how to do the wiring but it culminated with some unwanted heat and sparks. I then randomly tried various combinations of circuitry and discovered the concept of positive, negative and a complete circuit. Next came a brake light switch made from a simple clothespin. I fastened one leg of the clothespin to the handlebar. I put a loop of wire with a bare spot in it over the other leg. Press the free leg until a wire mounted on it made contact with the looped wire. Voila!... a brake light switch.

My first bike with five lights, speedometer, mud flaps, etc.

The Birth of the Hot Rod

The world benefitted from the technological advances gained during this massive conflict. The cessation of hostilities also freed up researchers so they could turn their efforts toward non-war endeavors like the basics of the computer, kidney dialysis and Velcro. Velcro may seem modern to some of us but was actually invented in 1948. By the early 1950's we went to the beach in the summers and played our fantastic little transistor radios.

Young men came back from the war, got civilian jobs and started families. Time and money were more available to spend on a new idea: hot rodding. The Model A Ford which was manufactured from 1928 to 1931 was ubiquitous. Guys liked the availability and the style of this affordable little car. The 1932 Ford had the first "flathead V8" for that company followed by many other Fords through 1953 that had this engine, which just begged to satisfy the young man to make his Model A go faster. Just put a hopped up '48 Ford flathead in a '31 A roadster or perhaps a later model overhead valve engine in an A coupe or sedan and you are ready to cruise!

The term "hot rod" was becoming more and more popular. It is very common now but when and where did it originate?

The term "hot" is synonymous with activity, action and motion. This is easy to perceive. "Rod" is a bit more inexact. Some say it refers to the engine's camshaft, which is a form of a rod with cams on it that opens and closes the valves. A hot cam would be one that raises the valves higher than stock and holds them open longer than stock as well. Thus, it is a hot camshaft or hot rod.

It could have originally been a contraction or variation of "hot roadster" since many of the early racing cars, especially in California, were early Fords and other makes that often raced as an open car or roadster on the dry lakes such as Bonneville in Utah and Muroc Dry Lake in California.

A simplification of the expression could be: A hot rod is an automobile that is rebuilt or modified for high speed and quick acceleration.

On a spring day in 1951 I first laid eyes on a hot rod Ford. I was a fourteen-year-old eighth grader at the local baseball field. Around the field was the running track for the University of Vermont track and field team. The high school track team shared the field with the university. A high school track meet that I had been watching had just been completed and I was waiting for several cars to pass on the exit road in front of me. Then it appeared. A high school senior and track man, Harmon Graves, came roaring by in his '30 Ford Model A roadster. It was not very much "rodded" by modern terms but without a hood and all four fenders cut with a swept back look and a roadster to boot it stood out as something very special to me. The engine was the original four-cylinder. I distinctly remember seeing the glass fuel bowl that was used in those days which would be frowned upon in today's safety consciousness. Hot rodding had reached me in little old Vermont.

Left: Harmon Graves and pal Ralph Clark having fun in their "A's."
Right: Harmon and his girlfriend out cruising the town.

My Magnificent Columbia Five Star Superb

I had to wait a few years before I could get my first car, let alone a hot rod so I busied myself with electric trains, making and flying kites and participating in sports. I had a paper route and decided to save money for a new bicycle. I saved up $85 and bought a Columbia Five Star Superb from a local toy store. Quite a bit of money for 1952. The bike had all the factory equipped "bells and whistles" such as a locking spring fork, a horn in the tank and a "carrier" with a brake light. I still have the bicycle today. It is very rusty from sitting in my father's barn for many years. The carrier disappeared somewhere along the way, so I found and purchased one just like it on the Internet. The new carrier made it complete with no missing parts.

My trusty (and rusty) 1952 Columbia Five Star Superb Bicycle

Some of the notable worldwide events that happened in the 50's were:
- Elizabeth II becomes the Queen of England after her father, George VI, dies.
- The Double Helix DNA Model is revealed by Francis Crick and James Watson.
- Sir Edmund Hillary and Tenzing Norgay become the first people to successfully climb to the top of Mount Everest, the tallest mountain in the world.
- Jonas Salk develops the first polio vaccine.

A '53 Plymouth four door sedan like my Dad's

In 1953, at the age of 16, I helped my dad buy a new car by looking through all the magazines we got in the mail and cutting out the advertisements for the modern automobiles and making up a scrapbook. Dad settled on a new '53 Plymouth Cranbrook four door sedan but he didn't quite go for my idea of "nosing and decking" it. I was learning the "lingo" of hot rodding but wouldn't have known just how to do the nose and deck job had he consented to it.

No Car of My Own Yet; I'll Make a Go-Kart!

I got my driver's license in 1954 but it would be another five years before I would buy my first car. Meanwhile I built a "Go Kart." I bought some Soap Box Derby wheels, built a platform out of fence boards and mounted an engine on it. It was a little Briggs and Stratton four-strokes-per-cycle engine with a horizontal crankshaft. The big challenge was to find a way to connect the engine's output shaft to the rear wheel or wheels.

At first, I attempted "direct drive" with a pulley on the engine's crankshaft snout and a pulley bolted to the rear left wheel. This was a good introduction to gearing as the high RPM of the drive pulley simply tore up the vee belt instead of driving the kart. I needed a gear reduction of some kind. I spent time thinking about transmissions and gear ratios and then came up with an idea that wouldn't cost me much. I found an old bicycle frame in the neighborhood, cut the front half away and mounted the rear half on the go kart. The idea was to run the vee belt from the engine output pulley to a large pulley mounted to the spokes of the bicycle wheel, then another small pulley mounted near the hub of the same bicycle wheel would drive a second belt to a slightly larger pulley on the Soap Box Derby drive wheel. There was lots of trial and error trying to keep the first belt tight and devising a slipping clutch on the second belt to act as a clutch, but amazingly enough the ratio turned out to be workable. Later I invested some money in ball bearing pillow blocks and machine shop work to utilize a final chain drive, and thus got rid of the big odd-looking bicycle frame. I was able to drive the little hot rod on the sidewalks and in the street where I lived. My street was a little "dead end" street with only eight houses and little traffic.

The local cops got after me only once when I ventured a little too far away from home. The cop just gave me a warning and said to limit my cruising to sidewalks near home.

A week before I went off to college in September of 1955, I called the local newspaper, The Burlington Free Press, to see if they would like to write a story about my motor-

ized Go Kart. A reporter came to the house and interviewed me and took pictures. On the very day that I went off to college the story appeared in the paper. How proud I was!

In 1958 I found a copy of Popular Mechanics magazine and found out how to make submissions to it. They accepted my write-up and a picture and my Go Kart I now called the "Stultz Special." In a few months the picture and article appeared in the magazine. Popular Mechanics was printed in several languages and distributed all over the world. I soon received a nice letter from a fellow in Brazil who wrote me in Spanish asking for plans for making the car. I sent him the plans and he answered with thanks and sent some mementos from his country as well.

Arthur Stultz Builds a Racer: Cost: $25 and a Lot of Work

Four years of college went by with very little progress developing my interest in cars. I lived on campus so had no need for a car, let alone a hot rod. The nearest I came to a hot rod was when I witnessed Dan Smith, a fellow a year behind me in college, being stopped by the town cops while driving his cool '34 Ford five-window coupe too fast on his way to class.

And Now a Motorbike

Back to my Columbia bicycle. I am sure that many of you guys out there reading this are saying you got your first car at a much younger age than I got mine. Probably

Homemade motorbike-summer of 1959.

so. I had to be satisfied at this time in my life with a Go Kart and bicycles; it simply was not my time to have my first car. I figured a motorbike might be better than walking for the summer of 1958, so I took off the carrier and rear fender of my bike and mounted a Briggs and Stratton engine in their place. This was the same B/S engine I had used for my Go Kart. I fastened a pulley from a Whizzer motorbike engine kit to the spokes. I concocted a drive system that used an automobile water pump pulley, plumbing pipes and long vee belt. I had made a motorbike!

Finally, My First Car!

I graduated from Springfield College in 1959 and was accepted to graduate school at the University of Vermont. Now I needed a car. I lived with my parents eight miles from campus so had to have wheels for the daily commute to my classes and back. Finally! An actual real live car of my own! The time had arrived when I would look for a good used car in town. I was supposed to buy a car for basic transportation to college and back but I was being bitten hard by the "hot rod bug." I knew I wanted a Ford. I looked at various Fords at several local used car lots and narrowed my choice to either a '53 Ford four-door or a '54 two-door sedan, both for sale at the same used car lot. I knew very little about the Ford engine designs and types of that era so didn't realize the '53 was a flathead and the '54 had the first of the OHV engines, so this fact didn't influence my choice. I had to agree with my buddies that the two-door was the way to go for a young single guy, whereas the four-door was for a stodgy old family man. I excitedly bought my first car, a 1954 Ford Mainline two-door sedan with the 239 cubic inch displacement overhead valve engine.

I was two months away from the start of graduate school so had some time to "nose and deck" my new possession. I had no welding skills so I filled in the hood and trunk lid holes with "liquid metal," did some sanding and sprayed on "rattle can" primer followed by matching color, also from a spray can.

My '54 in front of my house in Burlington, Vermont, just as I bought it.

I put lowering blocks on the rear. With the trunk lid lock gone I opened the lid by way of a manual cable control. I installed a "Bermuda buggy bell" that was operated by my left foot now and then. I finished off the customizing job by removing the stock grille in favor of a '48 Mercury grille I found in Maine. I bought a set of "spun aluminum" wheel covers downtown at Sears.

New Adventures in the U.S. Air Force

After I finished my one year of graduate school the Army draft was after me, so I inquired at the local military recruiting offices for "another way out" and joined the U.S. Air Force. I had been a science major at college so got stationed at Brooks Air Force Base in San Antonio, Texas, in the School of Aviation Medicine. I had landed in the heyday of space exploration. In October of 1957 Sputnik 1, the Russian satellite had circled the earth and challenged the United states to a space race. In 1961 President Kennedy responded with a challenge for his country "to land a man on the moon in this decade."

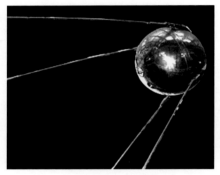

A replica of Sputnik

I was proud to be involved with the space program. Many of my Air Force buddies and I did medical research during Alan Shepard's and Gus Grissom's travels into space and John Glenn's orbiting of the earth. We tested Air Force pilots' reactions to stress and simulated weightlessness to show how man would react to the environment of outer space.

Projects Mercury and Gemini were followed by Apollo which aimed to fulfill President Kennedy's challenge to land a man on the moon.

I was excited and thrilled to be part of the space program but knew I would be out of the Air Force in the summer of 1964, so while serving my country I also pursued my interest in automotive, especially hot rods and customs. I enrolled in night classes at a junior college in town and studied for a degree in Automotive Technology. My trusty '54 Ford took me to the heart of San Antonio to little St. Philip's College two nights a week to study the various systems of the automobile: the engine, drive train, fuel, suspension, transmission, etc. I received an Associate of Arts degree in Automotive Technology in 1964.

When I went home on leave in 1961, I drove my '54 Ford back to the air base. I brought along the Soap Box Derby wheels and axles and the B/S engine. The air base had a wood hobby shop where I constructed a frame for an entirely new car. I even found another identical engine for it so had two engines with a centrifugal clutch on it

as well. I redubbed it "Art Kart" and even did some pinstriping on it. I didn't have a real good place to drive it but got permission to take a few spins around a local commercial Go Kart track. Time was up with the little machine in 1962 as I sold it to an Air Force captain who was stationed on the base. He had a young son who wanted it.

An Air Force friend of mine got discouraged with fighting his '49 Mercury which kept breaking down so I bought it from him for $35. (1963 prices...) I soon discovered it had a cracked block so I found another block, rebuilt it with knowledge I had accrued in my college engine course, and won a few trophies at the local drag strip on Saturday nights with it. The air base had an excellent automotive "hobby shop" for members of the military so I had everything I needed to keep the '54 and '49 going. I couldn't keep both cars very well, however, as it was getting expensive on an enlisted G.I.'s pay. I sold my '54 Ford to another Air Force guy. I did some basic "hopping up" of my '49 Merc, such as milling the heads and installing a ¾ camshaft (purchased from J.C. Whitney) an electric fuel pump and traction bars, and won a few trophies at the local San Antonio Drag Raceway. I had very little money so didn't do a thing with the body and interior but put what money I had into racing modifications.

A shabby lookin' '49...with a few trophies resting on the hood.

The drag strip had a newspaper that had a classified section. After one Saturday night of racing I picked up a copy of the paper and noticed there was a 1932 Ford coupe for sale. Wow! Like anyone interested in cars, I had read many magazines like *Hot Rod* and *Rod and Custom* and had decided long ago that my favorite hot rod was a '32 Ford five-window coupe. Now here was one for sale! I went to look at the car right away. It was completely disassembled and didn't have an engine.

The bare frame was out behind the owner's house, rusty and covered with weeds and small bushes. The body was in his cluttered garage. The interior was just a skeleton with no upholstery. A stack of wheels, paint cans, tools and other car paraphernalia resided quietly inside. The fenders and headlight bar hung against the wall. Running boards were in a corner.

Used by permission of the artist, George Trosley

This sorry sight might have been enough to discourage me, but like most intrepid street rodders-to-be I could envision a great future for this rough assemblage of metal, rubber and glass.

Myself with my '32 at the air base in 1964 and at home in Vermont in 2004

Rex, the owner, was asking $200 for it. I excitedly told him I would like to buy it and could pay him $30 each payday. He was a bit skeptical about my proposal of monthly payments but consented. I paid him loyally as promised and he let me work on it at his home every weekend. He had many car projects going on as well, and offered me a '58 Corvette 283 engine also for $200, so I kept paying and kept putting things together for the next seven months. I had it all paid for and somewhat put together in September of 1963. I towed it to the Air Force hobby shop where I continued to work on it to make it safe and drivable.

I was distracted a bit from my auto endeavors by a young lady in town. We got married in February of 1964 just a few months prior to my getting discharged from the Air Force.

Another adventure ensued as we moved from Texas to Maryland in the late summer of 1964. I had secured a job teaching high school Automotive Technology.

Everyone should do what they like to do as their life's vocation. Going to a job each day if it is nothing but detestable drudgery that you can barely tolerate makes for a very long day. There are too many hours in a working day to be stuck in something that is void of interest, challenge and fascination.

I had found a job that I really enjoyed. Working each day with young people teaching them about engines, toe-out-on-turns, differentials and, of course, throwing in an occasional lesson on hi-lift camshafts, milled heads and multiple carburetors was just "frosting on the cake." The job provided a great deal of satisfaction.

During my six years teaching in Maryland I did what I could with the '32. My wife had two young boys from a previous marriage, so raising them and another child of our own took precedence over putting in a set of Hedman headers or a Mustang II style front suspension in the hot rod Ford. I kept it on the road by simple maintenance and making low-cost modifications as needed. A plywood dash with ten simple toggle switches didn't cost much. I could dream at no charge.

I came across a junker '55 Chevy in a town nearby when out looking for a car for my stepson, Norman, who had just gotten his driver's license. He didn't particularly want a car without an engine or transmission. I thought it might make a good car for his younger brother so talked with the owner of the dealership and he simply gave it to me! What a deal!

My '55 Chevy 150 in 1991 before a seven-year total rebuild

Back Home to Vermont

In 1969 I heard that a new technical school was to open outside of my hometown back in Vermont. I thought it would be a great idea to move back home again. I applied for the job teaching Automotive Technology and got it. We moved to Vermont that June with an entourage of three cars, including the '55 pulling a trailer with the '32 on it.

In the new school in Vermont in the fall of 1970, things were looking up with better pay and the two stepsons setting out on their own so could afford to make upgrades to both the '32 Ford and the '55 Chevy. Oh yeah...I traded the high mileage family car for the '55 Chevy as my #2 stepson went off into the Navy. The car trade was good for both of us.

Some notable happenings in the 1970's were:
- NASA's Apollo 13 Moon Mission returns to Earth successfully after abandoning its mission to the Moon after experiencing oxygen tank problems and an explosion.
- The band "The Beatles" announce that they have disbanded.
- The Walt Disney World Resort is opened in Orlando, Florida.
- U.S. President Richard Nixon resigns from office after being implicated in the Watergate scandal.

I enjoyed teaching Automotive as much as ever as I continued with the improvement of both cars. I could use the tech center lifts after school and weekends which was quite an advantage.

In 1973 the movie *American Graffiti* premiered. Probably all or most diehard car guys have seen the movie more than once, as I have. The fact that the two main cars in the movie were like mine, a '32 Ford five-widow coupe and a '55 Chevy Model 150, whetted my appetite for continuing my quest to get my two cars looking as good as I could make them. I found a source for borrowing the movie and showed it to my students each year. I could get away with it as it featured cars, cars and more cars. I gave them a short quiz after the showings!

Street Rodding Arrives

I attended my first Rod Run in 1978 and "got wind" of the National Street Rod Association. The term "street rod" was being used more and more to signify a hot rod that was modified for safe driving on the street. A member of the host club of the rod run showed me a copy of the NSRA magazine StreetScene.

In 1983 I joined a local car club, the Champlain Valley Street Rodders. I was voted in as Secretary the following year and have served in that capacity ever since. Our club put on a "Rod Run" from 1981 to 1996 and have always done many car-related events such as club meetings, cruises, visits to local car related businesses and an annual banquet.

Some notable happenings in the 1980's were:
- Mount St. Helens erupts in Washington state.
- John Lennon is shot and dies.
- The popular video arcade game "Pac-Man" is released.
- The United States boycotts the 1980 Olympics in Moscow.

The 1980's found me doing lots of experimenting with my '32. There was straightforward improvement with the engine, suspension and interior but I also did some "far out" experimenting as well. I drove to a car show out of state with two working transmissions in the one car. There was the usual automatic Powerglide and behind it was a manual Chevy three-speed turned end for end. Running the manual in second gear gave me overdrive! A few years later I heard of a Cadillac engine that could run on half of its eight cylinders, so I put two carburetors on my small block Chevy engine, found a way to cut off the fuel delivery to one of the carbs and run on just the other one which served every other cylinder in the firing order. I went to one of the Nats East NSRA shows in this manner, swapping from eight cylinders to four now and then and saved a little fuel. Both of my modifications were interesting and enlightening but in the long run were not quite totally workable. Live and learn.

Some notable happenings in the 1990's were:
- Nelson Mandela is released from prison in South Africa and becomes the leader of the ANC.
- East and West Germany are reunited after the collapse of the Soviet Union.
- Margaret Thatcher resigns from her position as Prime Minister in the United Kingdom.
- One of the most complete T. Rex fossils is found in South Dakota and is named "Sue" after the paleontologist that discovered it.

Retirement...but Plenty to Do

I retired from teaching in 1991. This allowed more time to build and modify my two cars, especially the '55 Chevy. I got the wild idea to put a Ford engine in the Chevy which surely is not done very often. The school where I taught was given three 5 Liter Ford engines for teaching purposes, so I learned a great deal about these engines as the students and I worked on them. We mounted one on an engine stand and got it running. I went to an auto salvage yard nearby and found a similar '89 Mustang engine and put it into the car. It wasn't easy as there were many challenges. I also built a tilt front end whereby the entire front end consisting of fenders, hood, grille and bumper moved forward seven inches with an electric power seat motor and then I could tilt it up 55 degrees manually which gave me very good access to the engine for show purposes as well as easy access for working on the engine. This also proved to be quite a challenge but is a big hit at the car shows.

NSRA Comes to Town

It was a great day in 1993 when rodders in Northwest Vermont found out that NSRA was thinking seriously of presenting a show in the town of Essex Junction the next September. NSRA had many shows all over the country but it would be great to have the three-day show, which promised to have over 1000 cars or more in attendance just three miles from my front door. I helped to serve on the committee formed to organize the many aspects of the show and it indeed became one of the many popular NSRA shows held each year.

Some notable happenings in the 2000's were:
- The September 11 attacks. Four attacks by the Islamic terrorist group al-Qaeda against the United States.
- Hurricane Katrina. The very deadly and extremely destructive hurricane struck Florida, Louisiana and Texas.
- The tsunami in East Asia. An earthquake in the Indian Ocean created a tsunami that led to massive destruction and loss of life.
- The capture of Saddam Hussein by U.S. soldiers.

In 2002 I was asked to be the Vermont State Representative of the National Street Rod Association when Ken Bessette, another local rodder, moved up from this position to become Northeast Division Director.

The World of Racing and Car Shows

With the great upsurge in the interest of hot rods, customs and racing since the end of World War II, a number of Automobile Associations have sprung up across the United States.

Here are just a very few of the associations that promote and support the various types of racing and interests in automobiles in general:
- The National Hot Rod Association and the American Hot Rod Association are designed to offer safe and well controlled drag racing.
- There are hundreds of "roundy-round" race tracks across the country which present oval track racing. The United States Racing Association was formed with a mission of growing local and regional dirt track racing.
- NASCAR, the National Association for Stock Car Auto Racing, is the governing body for the extremely popular stock car racing that attracts thousands of attend-

ees at the huge races across the country.
- INDYCAR is an American based auto racing sanctioning body for Indy car racing and other disciplines of open wheel car racing. Most everyone has heard of and many people have seen in person or on television one or more of the Indianapolis 500 car races held on Memorial Day weekend every year.
- The Goodguys Rod and Custom Association promotes and produces hot rod, classic and truck events yearly.

The National Street Rod Association (NSRA) was formed in the late 1960's to promote an interest in pre-1949 street rods. In the last several years they have included cars thirty years old and older in the organization. As of 2018 they present ten weekend car shows covering all areas of the United States each year. The shows begin in April and end in September. There are seven NSRA divisions across the country with a Division Director for each division and one or more State Representatives for each state.

A very important feature of NSRA is their concern for safety. Each state has one or more safety inspectors that inspect street rods and customs on sixteen safety requirements as well as seven more recommendations concerning safety equipment.

NSRA also offers what is called "Fellow Pages" which is an emergency reference booklet that can be used to contact a registered participant should a person need help with a breakdown or other problem when they are traveling away from home.

Safe and Responsible Street Rodding

Street rodders buy, sell, trade and drive their cars and get extremely involved with the building of the vehicle from perhaps a totally stock old frame and body to just tinkering with a rod that someone else did the majority of the work on. The thousands of street rodders involved in the hobby have a great variety of skills when it comes to building a workable and safe car. The rodder may be a well-schooled and experienced mechanic and builder or someone who simply likes cars but has next to no knowledge of the skills needed to build and maintain his or her vehicle. Accidents and mishaps are bound to happen to the experienced or novice street rodder.

Street Rod Horror Stories

This book has been written to entertain the reader with some very interesting stories but also to promote the construction and maintenance of a specially built and modified vehicle to be safe, roadworthy and reliable. It is hoped that the reader will take each story as a lesson in what can happen if there is a lack of knowledge of what is safe and what is not, as well as just plain carelessness and unconcern.

I myself have had a few mishaps with my own rods and customs since I got my own car in 1959, and have taken an interest in what I did wrong or what was simply a stroke of bad luck or circumstance. I have tried to learn from mistakes I made in construction of my cars as well as how I drove them that caused an accident or mishap.

You will find ten of my stories here.

I hope you like reading the eighty stories and can also benefit by what the authors of each story experienced. Let's all appreciate our very interesting and enjoyable hobby of street rodding...*and let's be safe out there!*

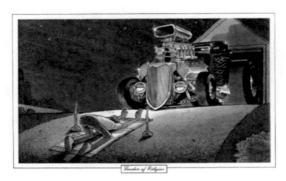

Used by permission of the artist, George Trosley

Bridging the Gaps
By Bob White
Cherryville, North Carolina

I have been interested with street rodding since I was a young teenager in the 1960's. I first joined the National Street Rod Association in June of 1992 at Nationals East in York, PA. Over 5,500 vehicles were registered for the show and had to be 1948 and older. I was driving a 1937 Chevrolet Streetrod pickup that I spent many years building from the frame up with help from my Dad and my Son and won many trophies over the years. National shows I drove to include Knoxville, TN, Columbus, OH, York, PA, Tampa, FL, Kalamazoo, MI, and Louisville, KY. accumulating over 150,000 miles on the 1937 Streetrod.

In August 2003 at the Kentucky Expo Center in Louisville, KY I was the lucky winner of the NSRA top participate prize a 1932 Ford Coupe build by Street Rods by Michaels. To be eligible for this prize you must be an NSRA member, have a Streetrod registered at the show on the Fairgrounds, and be present at the drawing. After winning the 1932 Ford I gave the 1937 Streetrod to my Son who still owns it.

I proudly drove this car to shows from 2003 until May 2011 when it was involved in an accident near Newport, TN on interstate 40 on my way back to North Carolina from an NSRA Show in Knoxville, TN. a 180 mile trip and the car was totaled.

The cause of the accident was due to rough roads, pacifically diagonal expansion joints at the beginning and the end of several bridges. The car was equipped with a straight solid round front axle with coilover suspension. This suspension design relies solely on a panhard bar to stabilize the axle from shifting sideways. When coming off the last bridge the panhard bar bracket broke off of the frame allowing the axle to shift to left side which caused total loss of steering. The car immediately turned left striking cable guard rails in the median strip which further damaged the front suspension and tore it from under the car. The car rode on top of the cable guard rails for 100 feet then came down with the cables tight on both sides of car and the doors could not be opened. Luckily no other vehicles were involved and I did not have a scratch. When the 32 came to a sudden stop I turned off the ignition switch, turned off main switch to the battery, and grabbed the fire extinguisher because the car was full of smoke. I then took a quick inventory of myself and didn't find any injuries. Then a man was banging on the car and trying to see inside to see if I was alright. I gave him a big smile and thumbs up! Then the car surrounded by dozens of people, mostly wearing streetrod T shirts and there were many familiar faces from the show. Many of them were talking to 911 on their cell phones as they tried to get the cables away from the doors so I could get out but the cables were too tight. The man that got to the car first went to his car and got tools to unhook the cable that was holding the drivers door shut way at the end of the cables.

When the cable sprang loose I opened the door and stepped out to dozens of cheering people, then I got hugs from a lot of people. Then the police, fire trucks, ambulance and roll-back showed up. When the EMT's were satisfied that I wasn't injured at all, the police asked me what happened. I told them I don't know. I hit a bad bump coming off of the bridge and the car turned left and hit the guard cables. Then the man that first at the scene talked to the state trooper. That is when I discovered he is a Trouble Shooting Expert with NASCAR Racing on his trip home to Charlotte, NC from a race. He told the police he was following me and admiring the car when it went out of control, suddenly turned left and hit the cables. This man said he's seen hundreds of wrecks on the race track but never anything like this and was afraid to look inside of the car when he got to it. We talked while the roll-back loaded my '32 Coupe and some scattered parts and found that on his trip to Charlotte he would pass less than half a mile from my house. He stayed with me for over two hours while the '32 was hauled back to the next exit and put in a garage where we loaded all my things out of the coupe into his Suburban and he took me 145 miles to my front door.

Within the next ten days appraisers totaled the car, I received a big check from Grundy Insurance, bought back the remains, hauled it back to my garage and ordered a new frame and other parts from Street Rods by Michael. My phone was real busy for the next several weeks getting calls from streetrod friends wanting to know about things.

Since then I had a bad heart attack, was in hospital for 21 days and a little blue eyed blond southern lady helped me recover and talked me into getting married (she had a bigger garage than I did).

The resurrection continues with the chassis completed and the body in primer ready for paint. Front suspension is from Pete and Jakes chrome hub to hub, drilled I-beam axle with conventional '32 buggy spring and half-rack steering. It will be on the road again in the summer of 2019. The resurrection of the '32 continues. The chassis has been completed and the body is in primer, ready for paint. The chromed hub to hub front suspension is from Pete and Jake's. It has a drilled I-beam front axle with conventional

I am checking out the piece of glass I got on my elbow while in the car trying to find my cell phone.

'32 buggy spring and half-rack steering. I am quite sure this tried-and-true front suspension design will be much safer than the suspension that caused the accident. It will be on the road again in the summer of 2019. It would have been very easy for me to give up street rodding after the horrendous accident but I am undeterred when it comes to rebuilding and once again driving my much-loved "Little Deuce Coupe." My heart has been in street rodding for many, many years so I will continue to enjoy my favorite hobby for many more to come.

Getting the Shaft
By Don Amundson
Auburn, Washington

In July of 2017 I attended a car show at the Des Moines Marina, Des Moines, Washington. As a National Street Rod Inspector, I had my "light bag" of inspection materials with me but it wasn't my intention to solicit inspections. I spotted Bob, one of our supporters who attends our Appreciation Day in Tacoma. I stopped to visit and he told me an incredible story about what had happened to him during the past month. After our Tacoma Appreciation Day, he decided to install rack and pinion steering into his 1953 Ford F-100 pickup truck. He went on the Internet to see if he could find a local shop that would do the job for him. He had also talked to some friends who recommended a shop in South Hill Puyallup, a small-town southeast of Tacoma, Washington. After getting information on the phone and taking the truck to the shop, he decided to have the work done there. The owner told him the conversion wasn't very difficult and he could have the truck back in less than two weeks. After a couple of weeks, the truck was ready so he arranged for his daughter to give him a ride to the shop to pick it up. His grandson was also with them. He checked out the work on the truck and it seemed okay so he and his grandson hopped in and headed for a cruise-in located in the Fircrest area of Tacoma. He drove down the hill from Puyallup and proceeded several more miles to reach South Tacoma Way, an avenue that starts in Tacoma and goes south about thirty miles.

Driving about 30 miles per hour he rounded a sweeping turn to the left. He then turned the wheel to straighten the truck but discovered immediately he had no steering control. He let off the gas. The truck hit a bump and veered off to the right. It then went left, crossed three lanes of traffic and bumped up against a street divider before Bob managed to get it stopped.

As this was happening, a Washington State trooper spotted the truck's irregular movements so turned on his cruiser's blue lights and pulled up behind him. When he asked Bob what seemed to be the problem and found that he was not impaired, they decided to check on what ailed the truck's steering. Bob opened up the hood and was shocked to see what it was. The shop had used round stock steel tubing, not a splined solid shaft, to mate the bottom of the column to the splined shaft of the steering rack. That was bad enough in itself but they also had welded a u-joint to the tubing apparently before checking whether the other end of the u-joint was close to the same size as the splined end of the rack. It wasn't; the joint was 3/16" larger. Going ahead with the bogus connection instead of getting the correct splined u-joint for the application, the installer made up some wedge shims and drove them between the tubing and the splined r and p shaft. If they had tack welded the shims that might have worked but they didn't even do that. The trooper was so incensed that they would do such shoddy work that he told Bob to turn on his hazard flashers and they would go back to the street behind the truck to see if they could find the wedges. It took about fifteen minutes for them to find all of the shims. The trooper left Bob with the report of his findings along with what repercussions may have happened had he lost his steering somewhere else along the drive. Bob filed a claim against the shop that did the work. He took his truck to another shop that made it safe using "double D" shafting and the correct Borgeson u-joints.

Street rodders should, of course, build all aspects of their cars such as steering, brakes and suspension for safe, comfortable and reliable operation. There are many parts vendors that are more than happy to give the builder tips on proper construction procedures. The Borgeson Company and Flaming River, for instance, have instructions on their web sites for building splined or double D steering u-joints with matching shafting. They give instructions for proper measuring of distances as well as the rules for where a universal joint is necessary. Implementing these suggestions will assure the safest and smoothest operation of this very critical aspect of street rod driving.

Insulin Insult
By John Cordischi
Caldwell, Idaho

In 2014 I purchased a new black Chevy Camaro SS with a hot 426 cubic inch V8 engine and fancy Foose chrome wheels.

The next year I was driving home from the market and started feeling dizzy. I am a type two diabetic and take insulin. I must have taken too much of the medication before leaving home because I could not remember my way home and started driving in circles. Before I knew it, I was "out of it" as they say! When I "came to" I was driving on a straight road leading to Lake Lowell. I faintly observed a row of highway reflectors on metal stands beside the road. There were about thirty of them that my car proceeded to mow down. I couldn't take evasive action as I was apparently going into a diabetic coma. My car then veered to the left side of the roadway and ran into a three-hundred-year old oak tree which demolished the front of the car, as well as myself!

I awoke out of the coma facing two paramedics who were starting an IV in my left arm. I could taste warm blood trickling down my forehead. Thankfully, I survived. Thoughts flashed through my head of how I got to where I was. I figured that I had driven in a straight line from one side of town to the other, crossing a major state highway without running into another vehicle or causing injury to anyone. At the speed I was going I could surely have caused a fatality.

I wonder who was at the wheel that day! Believe me, the incident made me believe in another greater than myself and definitely convinced me to keep better track of my blood sugar readings and medications such as my insulin.

My new Camaro, much like me, was totaled but eventually repaired...and later won three car show trophies. They are reminders of both the powers that be and the powers of restoration, neither of which can be taken for granted.

A Six Day Trip
By Art Stultz
Colchester, Vermont

I bought my '32 Ford five-window coupe a year before I was discharged from the U.S. Air Force in August of 1964. I had a teaching job all lined up to start that fall so had to move the family from San Antonio, Texas, to Maryland. My then wife and two stepsons, aged 11 and 13, moved over to her mother's home in town since our household goods were all packed up and sent to Maryland by moving van.

I stiff-hitched the '32 to the family car, a '63 dodge Dart four-door sedan with the 170 C.I. slant six engine. I test drove the setup around the block and it didn't track work a dang. A week or so before I had caught a glimpse in town of a homemade tandem wheel trailer that had been built to haul a '34 Ford drag race car, so I found it again and bought it for $55. It seemed like a trailer was a much better idea than a stiff hitch. The Dart didn't have a trailer hitch so I bought one and put it on. I drove the '32 up onto the trailer. Standing back, I could see that the trailer tires were small and besides that were pretty old and cracked. Not having much money, I rounded up some used tires that were bigger and in better shape than the others. I was all enthused about this setup but when I put the first wheel onto the lugs and tightened up the lug nuts, the tire sidewall rubbed heavily on the frame of the trailer. I tried turning them around and due to the wheel offset the tires didn't hit anymore. I knew the chamfer of the mounting holes was now "backwards" regarding how the taper of the lugs was built but it seemed to work okay.

All the work took considerable time out in the hot Texas sun in August, but the neighbor lady kept bringing me lemonade which helped considerably. By now I could see that the weight of the '32 caused the trailer hitch to almost drag on the ground. Not good. I went out and bought a hitch that clamped to the bumper. This higher location of the ball was better but caused the bumper to twist with the weight quite a bit. I jumped gently up and down on the trailer tongue and it seemed to be okay, although it twisted too much to my liking.

The Dodge was still pretty low in the back so I took the '32 off the trailer and backed it on which put the engine in the rear. Although this setup wasn't "kosher" it made the weight distribution much better.

Off we headed the next morning on the 1700-mile trip to Maryland. I put a sign on the back of this assemblage that read "slow moving" as I knew we wouldn't be going very fast through the nine states to our destination.

We hadn't gone but 100 miles or so when we had a flat on one of the trailer tires. We limped into the next gas station and had a mechanic repair the tire. What with that incident plus the slow travel speed, we didn't make it out of the big state of Texas the first day. Somewhere before we hit the border of Louisiana it got pretty dark so we pulled off the road at a little turnoff, and with the aid of flashlights unloaded sleeping bags and slept next to the car and trailer. About an hour later we were awakened by a train whistle. We were all confused as we didn't know how close the train was. It seemed like we were right on the tracks with the train coming right at us! Our oldest son Norman bolted out of his sleeping bag and attempted to run...who knows where... so we had to grab him and settle him down as the train roared by us on the tracks perhaps twenty feet away. We all got calmed down enough to figure out that we were actually in a safe place after all and not as close to the tracks as we thought.

Back on the road the next morning I sensed the trailer wasn't towing just right. Yep...another flat. Got that fixed.

Somewhere in Mississippi toward the end of the day, I noticed in the rear-view mirror that the left rear wheel of the trailer had extreme negative camber. We continued on to a motel (one night in sleeping bags alongside the road was enough!) where I immediately saw that the rear trailer axle was broken! The trailer was built with two transversely located three-inch diameter pipes with vertically located thick steel plates welded on the ends and Ford spindles welded to the plates. The two pipes were thus "dead axles." With one of them cracked clean through it accounted for the negative wheel amber. I spent the evening in the motel parking lot coming up with a fix. I happened to have lots of odds and ends of tools and parts with me in the trunk of the '32. I had a foot-long piece of angle iron that I wired to the bottom of the broken axle spanning the crack, and used a bottle jack and wood blocks to raise the axle back up to just above level. When I then took the bottle jack away it settled down to just about level. Good to go!

The rest of the trip was pretty uneventful except for restless, bored young sons punching each other in the back seat or arguing about one occupying the other's sitting room. We got lost briefly in Georgia and had a hard time getting around and through Washington, DC, but finally pulled into our final destination, Denton, Maryland, on the Eastern Shore. A six-day trip! The little slant six engine performed admirably through-

out the adventure. The rear bumper bent but didn't break. The vocational director in Denton had secured a rental house for us which we easily found in the small town. In another week we were settled in and I started my career teaching Automotive Technology. When the time was right, I could tell my students about my harrowing trip from Texas to Maryland. What a horror story indeed!

Clearly the Wrong Tubing
By Ken Bessette
Williston, Vermont

Since I was a volunteer firefighter with the Williston Vermont Fire Department for over 25 years, I thought it would be fitting to build a street rod on the theme of a small fire truck such as a pumper. I built it with the frame and body of a '36 Ford sedan and grafted on the cab of a '47 Ford pickup truck, which took quite a bit of fabrication. The bed was built mainly with sheet metal to resemble a pumper equipped with hoses, pressure gauges and an emergency axe. The engine was the popular small block Chevy and the transmission a Turbo-Hydramatic 350. Of course, I had it painted red! I also put "Pumper 1" in gold leaf on the doors and "Last Chance Hose Co." on the sides of the hood. There was also some gold leaf striping on the fenders. It made for quite a popular display at car shows; lots of people looked it over with great interest and asked many questions. It was even featured in Street Rodder magazine in 1993.

I drove it to York, Pennsylvania in June of 2004 with Ray, a friend of mine who helped me with the driving. While we were cruising through the Pocono Mountains of Pennsylvania, I noticed that the transmission was slipping now and then. We stopped and checked the fluid level. It was low so I added some fluid that I had with me. At the next exit I picked up some more fluid in anticipation of needing it. We got to York safely and settled into our motel.

The next morning, we went on to the show. There was a "Street Rod Repair Shop" on the grounds which offered anyone tools and expertise with most any common repair job that they should need. We went there and examined the transmission and found a pinhole leak in one of the tubes that brings fluid to and from the bottom of the radiator for cooling. A fellow who was on duty as an assistant cut the leaking tube at the pinhole with a small tubing cutter. I found a short section of flexible tubing in my tool box and gave it to the fellow for inserting into the gap where the pinhole had been. I was the NSRA Northeast Division Director at the time so had many duties to perform on the grounds so the fellow did the job for me. Later when we returned, we added some fluid to the proper level, I thanked him and we went on our way.

At the end of the day we proceeded to go to dinner and then to the motel. Ray was driving and exclaimed that the transmission was starting to slip again. We pulled into a gas station and noticed a puddle of fluid in the same area under the truck as before.

Ed Miller, a street rod friend of ours, had been several cars behind us so we called him on my cell phone and asked for his help. We put one side of the truck up onto the curbing for better access. Ed got underneath and exclaimed that the replacement tubing was the clear plastic type and not the tubing that I had handed the fellow back at the fairgrounds. Being plastic, it hadn't lasted long with the heat and pressure and had burst. Luckily, I had some more of the correct tubing so it was a simple job to replace it.

The next day we went back to the repair shop and found the same fellow who apparently had taken it upon himself to use what he thought was better tubing than what I had handed him. I gave him a little lesson on the proper tubing for the job!

Gasoline and Droplights Don't Mix!
By Paul Benoit
East Hartford, Connecticut

My wife Sally is the National Street Rod Association Connecticut State Representative so has a great love of cars and especially street rods, as I do.

We had been to 41 straight Street Rod Nationals and were preparing to attend #42 on July 30, 2011 when disaster struck.

On this trip to the Nationals we were going to drive our '35 Ford coupe. As is often the case I was doing last minute maintenance and checks on the car in my garage at 1 AM on a Saturday night even though we planned to leave sometime the next morning. I thought it would be a good idea to change the fuel filter. I got under the car and clamped off the rubber hose that was connected to the filter but upon making the disconnect some drops of fuel found their way to the floor creating some fumes that with the aid of high humidity that night hovered just above the floor. I was using a common drop light with an incandescent bulb to see and accidently dropped it while moving around.

It shattered! Flames formed instantly and just as quickly headed for the walls of the shop as I scrambled as quickly as possible from under the '35.

There were fire extinguishers in the garage but with the pain from the flames that had scorched my face my thoughts were pretty disorganized. Sally, who was in bed, awoke to my screams and my panic. Light from the fire flickering brightly on the walls of the bedroom told her what had happened so she immediately called the fire department.

In mere minutes fire equipment, ambulances etc. arrived and got right to work on the blaze which was increasing rapidly by the minute.

I was taken to the hospital where I spent the night. I received the care I needed and was able to return home the next day to inspect the rubble.

Sally and I have been members of the Connecticut Street Rod Association for many years so 40 guys from this club as well as from another nearby club, The Tyrods, showed up the next morning for the ensuing massive cleanup. These guys worked extremely hard loading up three or four big dumpsters with debris. One dumpster was used just for the immense amount of metal that had accumulated in the shop over the years.

By the end of Sunday only the footprint of the shop remained. Destroyed in the fire were the '35 Ford I was working on, a '63 Ford Galaxie, A Ford T bucket roadster, and a '37 Ford sedan that was 2/3 completed.

Diehards that we are, believe it or not, we were still determined to leave for the Nationals. Because we were without "wheels" a few more friends came to the rescue and offered their cars for us to drive to Louisville. An old friend, Dick Coleman offered his '62 Chevy wagon so we loaded up and "hit the road". The trip was flawless. Upon arriving at the show, we were amazed that everyone was already aware of our disaster. I was also amazed to see so many station wagons present. (Remember when you couldn't give away a wagon?) Just for fun we entered the wagon in the Northeast Safety Pick Division of the competition for outstanding cars. To my surprise the car was chosen and we ended up in the Sunday "Winners Circle" surrounded by big dollar cars. (Probably the first and last time ever!)

Driving home was a bit depressing knowing that we had to face a big pile of rubble and start planning a rebuild of the shop which is the other half of this horror story. Just the shop rebuild permit alone was grief enough as it took months to obtain one. About seven months after initiating the permit process we were finally able to jump most of the hurdles before the first board could go up. The garage rebuild went smoothly although the bureaucracy process could reduce a sane person to fits.

In spite of all this grief, today we have a nice new garage, the '35 is being fixed and we are driving a '40 Chevy coupe and a '63 Ford Galaxie convertible. We were able to maintain our record of attending every Street Rod Nationals. Meanwhile keep your fire extinguishers in plain sight and know how to use them!

Off to a Bad Start
By John Coolidge
Colchester, Vermont

In October of 2016 I flew down to Atlanta, Georgia from my home town in Vermont. I rented a car to bring my son back to Vermont for a visit. I didn't have enough time to drive down and back in my own car.

I am not a street rodder but I like cars and appreciate a nice street rod when I see one. We were on the big and very busy Interstate 287 in northern New Jersey when we came up on a flatbed truck with a "potential street rod" on the back. I don't know much about street rods but it must have been a late 40s or early 50s car. We followed along for quite a while and my mind went to daydreaming about the owner and what he had in mind for the car. Perhaps he had just bought it and was all excited about the long-term build that surely was ahead of him at home, wherever that was. The car was a bit rusty but looked pretty solid so perhaps he could see the potential in it and was taking it home for a total rebuild. No doubt he was doing some daydreaming too about being pretty happy with his new acquisition.

The driver's mind must have been jerked into reality about then as his truck began to weave and lurch left and right. Something was definitely wrong with the steering or suspension as this fellow was surely commencing on a wild ride. All the other cars that were nearby seemed to catch the drift, literally, and we all slowed down and gave him a wide berth to get control of his rig and not crash. Probably all of us witnesses were holding our breath and hoping for the best not only for him but for ourselves in our cars as well.

Within a few seconds the rear end of the truck seemed to "explode." The left side wheel still on its axle came out of its housing and proceeded to spin and careen along the left side of the road and onto the median out of harm's way. The right rear wheel tore right through the fender and rolled down the highway in front of us! As I pulled to the right into the breakdown lane the wheel veered to the right, slammed into the guard rail and flopped down in front of me. I was able to stop just in time to avoid hitting it.

The truck slammed violently to the road. It fishtailing and careened until the driver got control enough to veer to his left and onto the median and come to a reasonably good stop.

I got going again and the other cars all sped up when we sensed the all-clear. I could see the truck was safely on the median and out of the way of all the traffic

That was surely a "horror story" for that poor guy to look back on and he hadn't even gotten his street rod home yet, let alone running and on the road!

18-Wheeler to the Rescue
By Dennis O'Brien
Charleton, Massachusetts

About 1982 I was driving alone in my '34 Ford panel delivery truck on an Interstate highway through busy downtown Hartford, Connecticut. The engine in this street rod was a 383 cubic inch Dodge. I could smell gas so I turned off at the nearest exit and as I approached a parking lot where I could pull off, flames started coming through the hood louvers!

When I got the truck parked, I jumped out of the cab and opened a side panel. I attempted to put out the fire with my fire extinguisher but it was so small it didn't last long and was pretty useless.

Thinking that the truck would burn to the ground I started to pull stuff out of the back to save as much of it as I could. An "18 wheeler" tractor trailer pulled up and the driver hopped out with a big five-pound fire extinguisher. He gave it to me so I was able to put the fire out. It made quite a mess, but the fire was out and the truck was saved. There was so much damage I was not able to pinpoint just what had caused the fire but it certainly originated with the carburetor or the feed to it.

I called Paul Benoit, a good street rod friend of mine who lived in nearby East Hartford. He said to bring my truck over to his place. The truck driver, who was local, called his buddy who owned a ramp truck and had him come over. The ramp truck driver took me and my truck to Paul's big shop.

We soon moved the truck from Paul's to the shop of Bob Barry, an NSRA Safety Inspector, mechanic and painter where I left it for a long time while Bob repaired the engine and prepped and painted almost everything from the cowl forward.

This was a good lesson in making sure the fuel system is in good order and to not only have a fire extinguisher in your vehicle but to make sure it is of large enough capacity for snuffing out any fire that you might have. It might be handy to have and accommodating 18-wheeler truck driver nearby on all occasions as well!

A Stinky Situation
By Bill Lyon
Hooksett, New Hampshire

In 1988, there was an annual Northeast Street Rod Organization car show in Lake George, New York, that a group of us went to.

We left the show at its conclusion and were heading home to New Hampshire. I was leading several other cars in my group in my '32 Ford two door sedan. Lionel "Puddy" Paris, a long-time street rodder, was behind me, Joe Orlando was next and Ora George brought up the rear. It was dusk as we left a two-lane road and moved onto a divided highway that left New York State and headed into Vermont.

I saw a skunk approaching on my right. He was obviously about to cross in front of me so I swerved to the left but caught the skunk with my right rear tire. The skunk exploded a red and yellow spray into Puddy's grille...but left nothing on my car at all!

We all had CB radios in our cars, which gave Puddy a chance to yell at me...like all the way home!

Thank You Fellow Pages...Twice!
By Larry Anderson
Maidens, Virginia

I am the Virginia State Representative of the National Street Rod Association. I have owned a 1940 Chevrolet coupe since April 2000 and have put about 130,000 miles on it. The Chevy has been in 26 states and one province of Canada. I have attended all of the current NSRA division events except Bakersfield, Oklahoma City, and Springfield.

My first on-the-road mechanical issue was in September of 2012. It happened on a trip from Chesapeake, Virginia to Kalamazoo, Michigan to attend the Street Rod Nationals North. After that event I participated in the Street Rodder Magazine Road Tour led by Jerry Dixie. The tour went through part of Canada and ended in Burlington, Vermont for the Northeast Street Rod Nationals. Our first stop on this over two-week trip was in Lewisburg, Virginia for a local car show. The night before the show my wife and I went to visit a street rodder friend near Lewisburg and on our way back to the hotel I noticed the transmission did not shift normally. The next morning as we drove to the show from our hotel, the transmission did not want to shift out of first gear. I checked the transmission fluid and while it wasn't on the add mark it was a little low so I added fluid, which seemed to help slightly.

After the show we headed to Pittsburgh, Pennsylvania to attend a baseball game on Sunday. This was about a four-hour trip. The transmission was not shifting smoothly, but once it shifted into overdrive the car ran great. I was able to baby it all the way to Pittsburgh, to our hotel which was next to PNC Park where the Pirates play. Fortunately, we didn't need to drive the car all day Sunday, the day of the game. I always carry the NSRA Fellow Pages booklet with me so was able to call a street rodder who lived nearby. He gave me the name of Army's Transmission, the shop that he uses. I called the place first thing Monday morning and they asked if I thought I could drive the ten miles to their shop. I told them I thought I could. Upon dropping the pan, they found bits of ground-up metal which no doubt came from the bushing for the rear planetary gear that was nearly disintegrated. The owner of the shop took us to a motel about two miles away and picked us up again after our transmission was rebuilt. They started the repair at 10 A.M. that day and finished about noon the next. We were back on the road by 1 P.M. The owner of Army's Transmission Shop treated us very fairly on the repair cost and the transmission has worked flawlessly ever since.

Another issue was in October of 2013. My wife and I had driven the coupe from Chesapeake, Virginia to Tampa, Florida for the Southeast Street Rod Nationals. After the car show we drove to Anna Maria Island, Florida for a vacation week on the Gulf Coast beach. Near the end of our week I noticed that the right front brake caliper was not fully releasing. I took the wheel and caliper off, cleaned everything really well and thought that I had solved the problem. After the week was over, we headed for Pine Mountain, Georgia for another week to tour the Callaway Gardens. During the visit I noticed that the caliper was not releasing again so I stopped at a local auto parts store to purchase a new caliper and brake fluid and planned to do the repair work in the hotel parking lot. After giving it some thought, I decided to check the Fellow Pages as I had done before. I found only one person listed for Pine Mountain. I called him and asked if I could come to his house and do the repair there in case I needed help with transportation to get more parts and he graciously welcomed me to his driveway. I pulled the wheel and removed the old caliper and installed a new one and connected the brake hose. I thought I was nearly finished, except for bleeding the brakes, when I noticed a fluid leak in the middle of the brake flex hose. The local parts store did not have the hose I needed but a place 40 miles away did. The fellow drove me to that store to get the hose. After about six hours I had fixed my brakes. The old hose was probably the cause of sticking brakes all along. To add insult to injury, not long after that a woman backed into our coupe with her car and put a small dent in the bumper.

Of course, over the years I have had a few minor issues with the car but nothing I couldn't repair myself without the help of the Fellow Pages. I have had my car safety inspected by the NSRA Vehicle Safety program almost every year since I have owned it.

Snowbanks Win Every Time
By Kevin O'Brien
Williston, Vermont

A fellow in my neighborhood in Burlington, Vermont, was well known as a thief and drunkard. Dave, we'll call him, was a pretty big guy. He had a pretty nice '69 Camaro but with all his shenanigans with the law he had lost his driver's license. That didn't stop him however; he kept on driving as is the case with lots of guys and gals who drink and drive.

Vermont is well known for having abundant snow in winter, so as usual the snow banks were pretty high as the snowplows tend to push the white stuff into big heaps all over town.

Big Dave was up to his usual drinking and driving one day on North Avenue, which is a well-traveled street north of town. He decided to pull off the street and into a big shopping center. In his drunken stupor he plowed into a huge snow bank and fell asleep with the engine still running, the transmission in gear and one of the rear wheels churning away!

A cop pulled up, hopped out of his car, drew his billy club, and rapped on the closed driver's window. Dave woke up. The cop asked him to roll down his window. He did. As the cop reached in to pull out the key from the ignition switch, he asked: "What the hell are you doing?" Dave glanced at the speedometer and said: "About 25…" Needless to say he was arrested.

My First "Street Rod"
By Art Stultz
Colchester, Vermont

The All-American Soap Box Derby originated in Ohio in 1934 and spread over the country as the years went by, and eventually all the states were involved with "gravity racing" for young boys up to age 15. The races came to my hometown in 1948 when I was eleven years old. I was just starting to get interested in cars but for some reason didn't sign up to be in the contest for the next several years. Finally, in my last year of eligibility, I went to the organizational meeting. They said we had to go out and find a sponsor and that may have discouraged me, so I never got a car built in time for the competition. That was a good lesson in time management!

I did attend the races as a spectator and took note of how the cars were built. I was thinking perhaps I could build a car with a set of the wheels that were common to all the racers. The wheels and simple straight axles were part of a package that contestants were required to buy for the cars. They were identical so that the contestants would be on an "even playing field" since the wheels were so important to the speed of the car.

I knew a fellow who had been in the races in the previous year, so contacted him and bought his set of wheels and axles for $5.00. I built a car similar to the soap box derby cars since I had the plans that entrants got when they signed up for the contest. I thought that perhaps I could pick up a gasoline engine somewhere and put it on the car and drive around the neighborhood. Of course, this was not an original idea as most every red-blooded kid who liked cars had this thought.

I built the car but was stymied as to where to get an engine. I knew it would be even more difficult to design and implement a drive system. To engineer a clutch, transmission, and good steering and brakes would be a challenge. While I did lots of thinking I simply pushed some of the neighborhood kids up and down the street in it which got old quickly.

I had two buddies my age, Ralph and Jerry, who one day in March proposed the idea that we go coasting down a hill somewhere. It wasn't too difficult to decide where as we had a big hill a short distance away from my home. The street was North Prospect Street which was almost level for a half mile or so but ended with an extremely steep portion of about 500 feet just before it crossed at right angles to Riverside Avenue and then continued on as Intervale Road. Upon further review we thought it would surely be a "suicide run" as Riverside Avenue was a very busy thoroughfare. I suppose one of us could have stopped traffic while the other two guys coasted down the big hill and crossed Riverside Avenue. Nope, not going to do it; much too dangerous. We thought some more about Intervale Road. It was not nearly as steep nor as busy. We knew that it was much less traveled so we might as well take our chances and perhaps we'd have a nice, exciting ride.

We walked the soap box derby car down the aforementioned steep North Prospect Street hill, crossed Riverside Avenue and continued maybe another fifty yards where the ground was almost level. The car was not very long so we had quite a time fitting the three of us onto it. It was my car so I was the driver, and Ralph and Jerry squeezed onto the boxy rear of the car and hung on as best they could.

A moment should be taken here to describe the construction of the steering and braking of this car. The Soap Box Derby organization was very clear with their descriptions of just how these two features should be built, as they were very concerned with the safety of the young boys in the races. Well before race day, contestants had to take their cars to a designated place where they were inspected to make sure everything was built according to the plans. I thought the way the steering and brakes were built was pretty decent, so I might as well build mine the same way even though I was never going to enter it into the derby races. I was pretty good with building things around my home with whatever leftover stuff was in the garage, down cellar or in the back yard. I tried not to spend money buying things when there was something similar around the house. I built my steering system with a simple wheel mounted on a long shaft with a cable wound around its center. The ends of the cable went through pulleys which were fastened to the heavy wood floorboards with "eyes" that screwed into the wood. I found some of the eyes on my dad's workbench so figured they were good enough to secure the pulleys. The cable ends then fastened to the respective ends of the front axle. The single brake was a simple heavy piece of wood that pivoted with hinges so that when the brake pedal was pushed a cable passing back to the piece of wood would pull it down and the rear end of it dragged on the ground and stopped the car.

Before getting underway we briefly discussed the roadway ahead and what we would encounter on Intervale Road as it got gradually steeper for the next half mile. Perhaps the greatest challenge was the railroad tracks. Yes, the railroad tracks! There was a set of rails that crossed Intervale Road at an angle, not straight across at ninety degrees. (We had no fear of a train coming along as we had never seen one in all the time we had lived in that area. It was an active railway, however.) The concern with the tracks was their angle to the road. I reasoned that I would have to steer the car far to the left and then cut it to the right at the precise time so that both front wheels would strike the first rail at the same time. Otherwise the right wheel would hit first and get jammed against it and then the other wheel would go another few inches and then hit the rail as well, which would result in the car taking a violent right turn and would probably tip over.

Okay…the guys knew how I was going to steer when I approached the tracks so we all got into, and onto, the car. The guy at the rear pushed off and we were on our way. The Soap Box Derby wheels were certainly made very well as the car picked up speed in no time. As the driver, I got pretty concerned in a brief few seconds about how fast we were accelerating and applied the brake, only to have the brake cable instantly break loose from the wood that was the drag. We gained more speed as we approached the railroad tracks without the brake. I followed the plan with my steering as we came upon the tracks and steered way to the left and then to the right to hit the tracks head on. My knowledge of physics was right on but my Soap Box Derby building skills left much to be desired, as one of the steering pulley eyes pulled right out of the wood floor board which led to extreme looseness so that my steering control was almost nonexistent.

I did my best to steer the car but with the steering cable being so loose I didn't hit the tracks as I had planned. Sure enough, the car took a violent right turn as it hit the first track, which pitched the car up onto its right side and threw the two guys behind me off onto the pavement. When the car came to a stop in a few feet I was still in the car, but was left holding it up with my scraped and bruised arm and hand. Since it was the month of March with the winter snow just about gone, my hand was in a mud puddle about two inches deep. I hollered to the guys to get the car tipped back up so I could get out, but they were too busy getting themselves taken care of as they had been sent sprawling onto the asphalt and were all bruised and scraped up, dancing around, and hollering in pain. They eventually calmed down enough to straighten up the car and get me out of it.

I looked the car over and saw that the right rear axle had been bent right where it changed from being square to the round part where the inner wheel bearing was. We could still push it home with the bent axle.

I later straightened out the bend in the axle and in the succeeding months managed to get a small Briggs and Stratton engine from a friend in Canada. I rigged up a drive system with half an old bicycle frame and various pulleys and gears and had a pretty respectable "go-cart racer" to drive up and down my short dead-end street. I made the local newspaper with it as well.

I kept the car in storage while I was in college and a year of the U.S. Air Force. Still having an interest in it, I then took the wheels, axles and engine to the air base in Texas where in my spare time I built a nice frame for it in the wood hobby shop. I got a second engine for it and then submitted a story and a picture to Popular Mechanics magazine. The magazine was printed in several languages and distributed the world over. One day I received a letter written in Spanish from a fellow in South America who saw the story in the magazine and requested plans to build one like it. I drew some plans and sent them to him. He was very grateful and sent me some souvenirs of South America to show his appreciation.

I figured that I had done all I could with it and was ready to move on to something else, so sold the car to an Air Force officer who gave it to his young son. Two years later I bought a 1932 Ford that in the early '70s would be dubbed a "Street Rod", as the term became very popular for street-driven hot rods. My Soap Box Derby racer was not a true street rod, of course, but the precursor to owning, making safe, and enjoying driving various street rods in the next 55 years.

My Soap Box Derby car, with a little room for Ralph and Jerry.

Battery and Assault
By Kevin O'Brien
Williston, Vermont

Battery Park is in Burlington, Vermont, on a bluff overlooking Lake Champlain. I grew up in this city and the park has always been a great place for picnics, bicycle rides and performances at the bandstand on warm summer nights. It affords an excellent view of the Fourth of July fireworks. The park was named for the artillery stationed there by American forces during the War of 1812. The kids like to climb on an old cannon from that era that is on display.

The park was also a cool place to show off our hot rods and custom cars (if we had them) back in my day of 1974. We would often meet there informally in the summer evenings to shoot the breeze and brag about our rides and who had the fastest set of wheels among those gathered there that particular day.

One evening I was there with my non-descript looking 1947 four-cylinder M-38 Jeep that I got at an auction. There were about 20 other guys there with their cars, trucks and other assorted vehicles. Two of the guys, Steve and Peter, started bragging about how fast their cars were and got into an exciting "brag-off." Of course, it culminated in a suggestion to settle the issue and go out somewhere and race. Steve had a Gremlin with a Camaro nose and a 396 cubic-inch "big block." Peter's car was a '65 Chevy Nova with 327 cubic inch, 350 Chevy power. Both cars were very worthy potential candidates for King of the Road that night! We discussed at length just how to pull this off.

The highway on the other side of town seemed to be the obvious place to race. Interstate 89 was built in 1959-60 to improve traveling from the northwestern part of Vermont to the central-eastern area. It passed right by Burlington so folks could get on and off conveniently.

The plan was completed. We cruised up to Exit 16 several miles away. Someone went ahead and marked off an estimated quarter mile. Somebody else (probably a young lady...I can't recall) acted as the starter to wave the beginning of the drag race. Since there were about 20 cars in our entourage, the two challengers led the way followed by what might be called a "blockade" that stopped to keep the regular Interstate traveling cars way in back from getting by and spoiling the fun. I volunteered to be dead last of all the blocker cars as my Jeep was pitifully slow.

This was all set up well in our minds but as we started forming the blockade with our stationary cars, we didn't plan on so many fast-moving regular drivers on the Interstate barreling down on us. We thought that they would simply see the blockade up ahead and casually slow down and stop behind us. But in no time, they were on us with a great screeching of tires and then they smashed, crashed, bumped, and careened into each other creating all sorts of havoc. A regular "Destruction Derby" for sure! An AMC

Pacer that was built with lots of glass left most of it on the highway. At the start of this disaster I could see the mayhem developing, so drove my Jeep over to the edge of the highway to get out of the line of fire. Meanwhile the two combatants had gone ahead with their race and were far on down the Interstate. The 20 blockers quickly assessed the situation and got the hell out of there as fast as they could, myself and my Jeep notwithstanding. Of course, the police, ambulances, and wrecker trucks soon arrived. The police spent several hours sorting out the mess and trying to find out what caused it all. Miraculously there was no one hurt in the melee. No arrests were made. The cops probably interviewed some of the drivers caught up in the massive destruction because they came to Battery Park several days later and started asking the owners of the hot rods and customs if they knew anything about the big accident. Of course, none of us knew a thing!

Jersey Barriers and Pennsylvania Roads
By Anonymous
Vermont

In June of 1984, a group of us cruised in our street rods from Vermont to the NSRA Nats East car show in York, Pennsylvania. I drove my small block Chevy powered '34 Plymouth coupe. The trip to the show and back took us through much of eastern Pennsylvania. There was a great deal of construction on both the interstates and other roads that summer. We encountered many alternating good and bad roads. The so called "Jersey Barriers" appeared now and then which offered just a narrow one-lane passage. There were lots of cracks in the pavement as well.

On the way back on that Sunday, my car hit some especially large holes in the concrete which caused the radiator support rods to flex. The radiator was then thrown back into the fan, which carved a quarter circle in the cooling tubes and fins and created a bad leak.

I pulled over to the side of the road and was followed by the several cars in our group. The obvious plan was to get the radiator fixed. In a short time, a state trooper stopped and used his radio to call one of the volunteers listed in my NSRA Fellow Pages book. Lucky for us, the fellow we called had a friend that owned a gas station at the next exit. We took off the radiator, loaded it into the trunk of one of the other cars and went to the gas station. What would be the chances of their having a way to repair it? We asked the fellow on duty about a possible repair and he said he was there just to sell gas on that Sunday and was not a mechanic. We spied an oxy-acetylene torch outfit nearby and asked if we could use it. The fellow was very accommodating and said we could. One of the guys in the group had repaired radiators in the past with an oxy-acetylene torch so we found some solder and he attempted the fix. He got it fixed at least visually but knew full well that under pressure it would probably leak. Looking around, lo and behold, there was a radiator tester! We put the pressure tester on the coolant filler neck, filled the radiator with water, plugged the inlet and outlet with rags and pumped up the tester. Sure enough, there were a couple of spurts of water escaping. We repeated the soldering and testing process several more times and finally got the leak stopped up.

We found a couple of empty gallon jugs, filled them with water and took them and the radiator back to the '34. We put the radiator back in, put in the water, made a second trip back to the station for more water, and proceeded on our way back to Vermont with no further troubles. We all drove a bit slower and more cautiously over those cracked and miserable Pennsylvania roads!

Flickers and Flames
By Al Downing
Syracuse, New York

In 1987, I had two street rods, a '27 Ford roadster and a '40 Chevy two door sedan. (The latter was my and my wife's "wedding car"!) I did lots of work on these two cars in my 24'x 26' garage. Like many garages in which we street rodders work, there were lots of parts here and there, as well as paint, solvents, etc. Both cars dripped a bit of oil and gasoline now and then. I spread so-called "kitty litter" to absorb the gas and oil.

One night about 11 PM my wife and I were getting ready for bed when I saw a flicker of light coming from the garage. I was pretty sure the building was on fire. The garage was a mere fifteen feet from the house. I immediately called the fire department and then rushed out to the garage but couldn't get in! I would, as a rule, bar the walk-in door from the inside, walk out one of the electric power rollup doors and close it with the remote. The fire had quickly spread and burned through enough of the wiring so that my remote wouldn't work to get in.

My good neighbor tried to get in as well, but looking back it was probably a good thing we couldn't get access, as there is no telling what would have happened next. Opening a big door would have let in lots of air which would have supported combustion even more.

The fire department with two trucks and lots of men was there in a short time as the station was close by. They broke down the door to get in. The T was on its wheels, so a heavy strap was fastened to it and several firemen pulled it out. I had been working on the Chevy and it was up on jack stands so there was no getting it out easily.

The fire department did a fantastic job in all respects. I can't thank them enough.

I had insurance with Sneed, Robinson and Gerber and they settled the claim quickly and efficiently which I appreciate.

I was able to rebuild both of the cars as well as the shop which sustained about thirty percent damage.

I am left with the question of just what caused the fire. I never found out exactly, but I have suspicions regarding the safety of the oil drying compound that I used. It was actual Kitty Litter and not Oil Dry, Speedy Dry or other compound that is made specifically for absorbing oil and fuel. Perhaps this will open some dialog among mechanics and street rodders about avoiding the use of Kitty Litter.

French Hill Fish Tale
By Dennis Caron
Colchester, Vermont

I have operated my own body shop in my hometown for many years. I have done lots of race car work in this shop for myself and others in the area. There are many oval race tracks in the northeast section of New England including Thunder Road in Barre, Vermont which is not far from home. In the '70s and '80s Catamount Raceway in Milton was only ten miles to the north of me.

During the 1987 racing season I built a race car based on a '87 Chevy Nova body and was all set to race it at Thunder Road at an upcoming race. I had purchased a trailer that was built to haul race cars. It was complete with a rack on the front of it to carry lots of spare wheels and tires and there was room for such necessities as a welder, floor jacks, safety stands, etc. The car fit pretty well but had to be placed toward the rear of the trailer, as there was stuff packed in ahead of it.

The race was on a Thursday night, so a fellow driver, Dave Whitcomb, and I set out on Route 2 on the way to Thunder Road with me driving and the trailer and racecar in tow. Along the way, in the town of Williston, we came to the well-known steep and winding part of Route 2 known as "French Hill." I knew the hill pretty well, so backed off the accelerator in preparation for the descent. The trailer started to fishtail. Dave and I couldn't decide if I should let off some more or accelerate to cure the fishtailing. I floored it! More fishtailing! Poor Dave's face was as white as a sheet! The trailer swung violently left and right as we careened down the hill. The road itself went to the left and then to the right again before it reached level ground. When we finally got to the level area, the swinging of the trailer threw off just about everything that was stored ahead of the race car. Wheels and tires flew off to the left, and the welder, jacks, a tool box and other odds and ends went off to the right into the "pucker brush" as they call it. There were no front yards of houses there at the time but lots of high grass, weeds and a swampy area where all this stuff went. I managed to bring the truck and trailer to a halt to the side of the road. We collected ourselves and then collected our equipment and put it back in the trailer, but the several wheels and tires that had flown off, were nowhere to be found. We could see 200 to 300 feet of black tire marks back on up the hill where we had done the fishtailing.

The race car was obviously placed too far to the rear on the trailer and this poor weight distribution surely caused the fishtailing. We thought about what to do next. Go to the race or not? Change the weight distribution somehow? I remembered that my brother-in-law, Ray, had a body shop just up the road in Richmond so we drove the rig (slowly) to his place. Once there, we took an oxy-acetylene torch and cut off the rack and the necessary bars, brackets etc. enough to be able to place the car more forward. This took considerable time but we managed to work pretty fast and off we went to the race track!

After all this, there was a silver lining to the whole black, cloudy mess! We not only made it to the track in time to race but we actually won the feature!

A strange but fortunate thing happened after that: back at French Hill some people who knew me and recognized my truck had been driving the other way on the other side of the road when I was coming to a stop with the truck and trailer and they found the wheels and tires we couldn't locate and returned them to me at my shop! A happy ending to a tale that could have been much worse.

A Tale with a "Nashty" Ending
By Allen Haines
Cumberland, Maryland

Often when the mood strikes me, I find myself driving hither and yon on secondary roads in the rural region that surrounds my hometown. I'm often surprised by the sights I come across. In many areas it's not nice to get rid of your broken and overworked vehicle. Removing the tags and the battery often suffice as the ceremonial farewell for a car or truck that has served you well for-however long. One man's junk, another man's treasure.

That was the background for the scenario I stumbled upon one cool spring morning. Rounding a curve, I spotted a dirty, seemingly neglected Nash Metropolitan stashed under a grove of trees. If its grimy appearance was any indication it had been there-or somewhere similar-for quite a while.

Seeing a fellow in the yard pushing a wheelbarrow gave me reason to stop. I was curious about the ol' Met, and perhaps he'd be willing to sell it to somebody who'd put it back on the road. Not me; I've got more than enough on my plate, but I'd be willing to shop it around to a fellow car-guy looking for something to build. And then there was that neat little primered example at the Street Rod Nats in Louisville last summer that got a lot of attention. I took a few photos of that V-8 powered beast myself, and I understand well the "new" challenges rodders are now taking on. This green and white example would fit the category perfectly.

Explaining my quest for vintage tin and pointing to his car, I began my spiel, only to be interrupted early on by his own story of search and rescue. First, the car wasn't for sale-at any price. He cautioned me to sit and be comfortable; this could take a while:

We have to go back some four and a half decades to begin this tale of mystery and intrigue. Mike Jr., my newly-found auto enthusiast, was a mere lad who relished the time he spent with his dad, whether it be fishing, picking strawberries on the farm, or cruising in this very car.

It was an afternoon excursion into town when the mystery developed: on a side street a few blocks from the main stem the car quit. His dad coasted to the curb, opened the hood and determined that the engine wasn't firing. Mike Jr. remembers watching his dad hold a screwdriver against a spark plug while he, Jr., cranked the starter. He didn't know the hows or whys, but he knew that Dad had found the problem and Dad could fix it. Dad could fix anything, he recalled.

A call from a phone booth to a friend got the pair back home, where, after dinner, the required parts and tools were slipped into a leather bag and the two Mikes, and Mom as the backup driver, left in the truck for the short journey downtown. Dad didn't think it should take too long to get the Met up and running; they'd be back before dark and all would be well. Wrong.

The car had vanished! There was absolutely no sign of its ever having been there. Mike Sr. was astonished; would it have been towed so soon? A visit to the police station brought no answers. He returned to the site and questioned, without success, the few merchants he could find; it was by then well after five o'clock and most stores had closed.

In a foul mood over the wasted time, Mike Sr. led his brood home and began calling all the towing companies in the area; they knew nothing. His prepared diatribe found no listener.

And so it would stay for the rest of his life. He heard nothing and saw nothing. Every time he spotted a Metropolitan that could be "his" he inspected it; there was a minor imperfection in the windshield, and Jr. took it upon himself also to conduct the same procedure each time the opportunity was presented, which wasn't often-even at the height of the lowly car's popularity.

Mike Sr. had passed on about 10 years previous to my visitation, no more knowledgeable about his car than on the day it disappeared. And Jr. had pretty much put it all to rest himself, concentrating on raising a family and, as time allowed, relishing his time spent fishing in the nearby pond where he and his father built their close relationship.

My questioning glance in the direction of the trees brought a weak grin from Mike. "I'm coming to that," he said and continued with his tale: "Nearly two years ago I got a phone call from somebody who told me I'd have to remove my stored car from the building he had just purchased. It would be demolished and a new structure would be erected.

"You can imagine my feelings at that revelation. I mumbled something unintelligible to my wife and took off for the address I was given. I was familiar with the location; I drove past the neighborhood often when I attended my son's school functions. It had served as a garage for a guy who performed home improvements, mostly painting and minor carpentry chores.

"I was shaking from head to toe when I opened the door of my pickup truck; was this another wild goose chase? If it's the real deal, how do I respond? I knew, when I stepped into a back room of the building, that I had indeed found the long-lost Metropolitan, even before wiping away a half-inch of dust and bird droppings in order to check the windshield flaw."

Mike stopped to catch his breath before beginning his wrap-up: It seems the man who made the phone call had no answers; he had bought the place with the belief that all inside was right and proper, that the car had belonged to the former owner, who had himself passed away recently. The surviving family members were of no help; they knew nothing, had never laid eyes on the little vehicle, now setting on flattened tires and surrounded in part by paint cans and other containers, up to 55-gallon drums.

The license plates were still affixed, and a readable registration card-with Mike's father's name listed as the owner-lay on the front seat. The rest was elementary, and Mike extracted the car he had enjoyed as a kid and hauled it to his home a few miles away. He'd build a garage to house it soon.

I stood, expecting to shake hands and part ways. Mike directed me back down. "There's more" he offered. "It gets better."

He had told his mother about the rescue, providing the details as best he could unravel them. When he mentioned the name of the late owner, now fingered as the thief who took the car and hid it in his business for so long, his mother stiffened, dropped her head and took a deep breath.

"I knew him" she began. "I dated him for a short while, but I never became as serious about him as he was about me. I met your father shortly after the breakup, and we knew we were meant for each other. He swept me off my feet, and we were married within a year. I had been told that the other fellow was bitter and blamed my husband for his loss. He promised that he'd even the score. I just never thought he'd go to this extreme. The scoundrel."

As I drove off, the saying wouldn't go away: "All's fair in love and war."

Do What Ya Gotta Do
By Roger Dussault
St. Albans, Vermont

Many of us street rodders have been in the position of rushing to finish our rod in time to go to a certain rod run or car show but the car really isn't quite finished. But close enough...we think.

In 1990 I had "finished" my '37 Chevy coupe just before my three buddies and I went to York, PA to the NSRA Street Rod Nationals East Car Show. I had put maybe 30 miles on the car and it seemed okay.

Street Rod Horror Stories

Buddies Paul, Gerry, Alan and I decided to put our four cars on a big "ramp trailer" behind Alan's big 1-ton Chevy four-door truck to haul our cars to the show. Alan's '32 Ford "Vicky" was also a "fresh car" and not all that reliable so that was a pretty good idea. Plus, we saved some gas by not having to drive all four cars.

The York show, as usual, was very exciting with lots of cruising around the streets near the fairgrounds, especially in the evenings. Route 30 was always busy with street rods everywhere going to restaurants and "dragging" between traffic lights. The host hotel where we registered was a site of activity also as the parking lot was one big car show in of itself.

The show didn't open until Friday, but lots of cars had been registered on Thursday so that night was a great time to cruise around town in our rods. We did our part that evening to escalate the excitement. We figured we would get separated so we all had radios for communication. Gerry had a wrecker business so was able to provide us with Motorola portable radios with our own dedicated frequency. No cell phones in those days. Worked great to call back and forth if the cops were nearby or if we simply wanted to generally communicate.

It's always been easy to get me "revved up." It took maybe two people to "egg me on" to spin my tires and do some "fishtailing" when the situation was right. I was showing off to the crowds here and there when a great cloud of burning something-or-other emanated from under the car. I hightailed it to the nearby hotel parking lot to check things out. The other guys had separated from me at that point, so I called all three of them on my radio and they came to join me.

I checked my oil and transmission fluid levels and found the transmission fluid was down quite a bit. "Was there a leak? What to do now?" We had limited tools and no really good jacks or other ways to lift the car so I could get underneath to see if I could repair what probably would be a leak somewhere. Hopefully it would be easily fixed.

I spied a manhole cover right there in the parking lot. The cover had "Municipal Sewer" in raised letters on it. Hmmm. We had tools to pry the cover off to reveal a sturdy permanent ladder that led downward.

I got down into the manhole and the other guys pushed my car over me. Not the greatest working conditions, but I managed. The guys handed me a flashlight and some tools and I tightened up a very loose clamp on a fitting for a tube that led transmission fluid to the cooler at the bottom of the radiator. Must have overlooked it before we left on the trip. Leak cured. While I was at it, I checked and tightened everything else within reach that I could. Should have done this before we left home.

As I was finishing up the guys started kidding me that they were going to leave me there in the manhole! Nice guys! Naw, they pushed the car back off from over me and we got more tranny fluid and were good to go. Sure, I got picked on and kidded a lot, but "ya gotta do what ya gotta do!"

The Angry Trailer
By Walt Fuller
Newnan, Georgia

In 1998, my wife and I and our good friends, Charlie Cox and his wife, drove from our homes in Georgia to attend the Frog Follies Car Show in Evansville, Indiana. I pulled a luggage trailer behind my '40 Chevy two-door sedan. Charlie drove his '46 Chevy Suburban.

We set up at the fairgrounds in the late afternoon on that Friday. Around 8:30 the next morning we decided to move to another site, which looked like a better spot to park the two cars and the trailer. The trailer was disconnected from my car with the top held up with a prop rod. As Charlie and I picked up the tongue of the trailer to make the move, I stepped in a hole in the rough ground and sprained my ankle! The momentum of the trailer moving slightly downhill caused one of the tires to run over Charlie's foot and the lid of the trailer to slam down and latch itself, while at the same time catching Charlie's hand between the lid and the trailer body and tearing a fingernail! The prop of the lid was simply not well constructed.

We got his hand out of the gap between lid and trailer and went to a first aid station where they fixed us up enough so that we could continue with the show, albeit we were both hurting enough to agree that this was a most unenjoyable weekend.

At home I made a more substantial prop rod that was well secured on both ends when the lid was in the up position.

A while later I wrote up the story and submitted it to my car club's newsletter. It drew quite a bit of attention from my buddies! Someone somehow got a hold of the story and submitted it to American Rodder magazine.

Such notoriety for simply not shutting the trailer lid before the move, which I certainly should have done. This is a good example of how an injury can come from not preparing a car-or in this case a trailer-for a move, even if it is a short one. Accidents and injury can happen in a brief amount of time and in a very short distance. Be prepared no matter what is a good motto!

Reaching High
By Bob Brunette
Fletcher, Vermont

Anyone who acquires and begins to use a "new" piece of equipment certainly knows there is a learning curve to enjoying the benefits of that tool or equipment. This is the story of my learning curve with my newly acquired, worn out, air over hydraulic automobile lift. Getting started on the repairs entailed replacing the two outer hydraulic tubes as they had holes in them. I did not want to trust welding them as they would be buried under the cement floor of the garage. Once that was finished and reassembled, I attended to the other lesser repairs and upgrades like piping, repairs to the valves, driving the oil pipelines under the concrete floor, cleaning the ramps, and assembling an air compressor out of various and sundry parts collected from here and there. All went well and it was finally beginning to come together as my "new" piece of equipment. I was proud!

First use of my "new" lift was to do some undercarriage work on my '46 Ford street rod pick-up truck. I drove it on the ramps and raised the truck to a comfortable working height, making sure I had sufficient clearance between the truck roof and the ceiling and the roof support beam.

During lunch that day Gloria and I made a decision to drive to Maine over the weekend to visit some relatives. We planned to leave that afternoon, as that would give us a little extra time for visiting. Before leaving, I went down to the garage to make sure all was well with the pickup on the lift and that all the safety devices were in place.

Monday morning, upon entering the garage, it seemed the truck was much closer to the support beam than it was when I left three days ago. A closer look confirmed the incredible fact that the front portion of the cab roof was tight against the support beam, leaving a neat 3" to 4" creased dent across the leading edge of the roofline. The interior overhead console of the truck I had made out of some 60-year-old tiger maple that a good friend had given me was also severely damaged.

"How could this happen? How could this be?" I was crestfallen. I had to regain some of my senses. I went into engineering mode to try to wrap my head around the situation. It took some thinking, but I finally realized what had happened. My first mistake was I had left the compressor on over the weekend and the pressure switches would respond to low or high pressure. With the inevitable slow air leaks in the air system, over time, the low-pressure switch would call for more pressure. Under maximum compressor shut off pressure, the lift would rise, lifting the truck incrementally over time and eventually into contact with the beam.

It took some time and effort to "jack" the crease out of the roof with my power pac, do the finishing work to prepare for paint, and then salvage the overhead console.

Lesson learned: shut the compressor off when leaving the garage!!!

A Dip in the Pool
By Mark Gissendaner
Valencia, Pennsylvania

I build and modify street rods for a living. Some jobs, of course, take many months or even years to build, so when they are done, I take great pride and satisfaction in the final test drive and then turning the rod over to the owner.

In 2012 I had finished one of the long-term builds of a '41 Ford coupe and called the customer to have him come over and get it.

He arrived with a car-hauling trailer behind his truck and prepared to load up. He drove it up onto the trailer, shut off the engine and stepped out of his newly completed "pride and joy." Unfortunately, he had inadvertently left the transmission in neutral and before he could get the wheel chocks placed properly behind the wheels, the car started moving slowly backwards from whence it came!

My street, where his rig was parked, goes slightly downhill so when the driverless car got up momentum from the descent off the trailer there was no stopping it. We raced after it on foot but only to warn people to get out of the way. It continued cruising down the street backwards for a half dozen seconds but then it veered to the left, jumped the curb, crossed the neighbor's lawn, crashed through his fence and landed fully into his swimming pool! Aargh! What a calamity to say the least!

The home owner was home at the time, so after much explanation and apologizing, we called a tow truck that winched the car out.

The car owner's insurance paid for a new fence, swimming pool vinyl liner and landscaping.

The car was immediately returned to my shop. The fluids of the engine, transmission and rear end were changed. The upholstery was totally replaced. The body and paint were not at all damaged. After two months, the owner drove it back home. *He did not trailer it!*

Difficult Decisions
By Allen Haines
Cumberland, Maryland

A rodder announced a forthcoming addition to his family. We knew he had a line on something special but he was tightlipped, promising we'd know in time. When the call came, we weren't prepared for the real deal. Sydney offered coffee, doughnuts and idle chatter while awaiting everyone's arrival. Neither he nor his wife Celeste would provide info about their latest purchase; they just smiled grimly.

When all were present, we went to their garage. Inside, a flip of the light switch revealed – not much. A tarp covered something and we saw the bottoms of four aged whitewall tires. Silence reigned and glistening beads of sweat formed on our upper lips. The suspense was chilling-would the unveiling be anticlimactic? Our minds raced, imagining what we'd see in just a few seconds. Wordlessly Sydney and Celeste took their positions and whisked the cover away, revealing a 1936 Ford coupe. Sydney's demeanor matched that of Celeste's: Uneasy smiles. The coupe – a dull red – was obviously a relic. We kept our distance, until Sydney urged us to come closer and look it over.

"This isn't local," I believed. It had an early Olds engine, a 303 it turned out, with a J-2 intake with three Rochesters. The starter was on the passenger side, relocated; a six-volt battery resided in the firewall. A Ford slave cylinder operated the clutch and a Cad-LaSalle top loader sent the ponies to a '40 Pontiac rear end, a 4.30:1 ratio from a six cylinder-all according to Sydney. A second sweep brought more details: Foxcraft skirts made the car seem lower, longer shackles made it so. 1949 Plymouth bumpers added style. Fiesta hubcaps covered the front wheels, no need for any in back. Inside, a red metalflake wheel sat atop the stock column, a horn button clamped to the side. A '49 or '50 Ford speedometer looked neat. A triple gauge panel was fitted under the dash. Bucket seats (VW?) awaited covering.

Then, the kicker: Somebody mildly chopped the top but never completed the job. A little metal finishing would wrap it up. We were seeing real history. What do you do with something like this? Back inside, Sydney had an important decision to make. A lady in a neighboring state sold this car, and she didn't want to see it again–no pictures, no visits. It had belonged to her husband's brother, who began to "hot rod" it in the early '50s. His daily driver, he was very proud of it, the "fastest car in town." Before leaving for Korea, he put it on blocks to await his return. That didn't happen. In the late '50s the coupe was handed to the lady's son, who vowed to finish it to honor his uncle.

The youngster, however, enlisted and 'Nam took him overseas, but not before he chopped the car's top; he'd complete it (incentive) when he returned – but he didn't. His dad was devastated, having lost a brother and a son. He couldn't touch the Ford, but Dad spent time alone with it lost in thought. His wife would often find him–when it was too silent–sitting in the garage, handkerchief in hand.

He asked his wife to keep it as is, as he lay dying. And she tried, the best she could. But the garage was showing its age, as was she. The time had come, but she couldn't bear the thought of someone else driving it. She'd sell the car, reasonably – the farther away the better, bringing everything up to date. "So now," Sydney asked, "what should be done with it?" Occasionally, we have the opportunity not to make a decision. This was that time. I said "Good-bye," expressed my support for whatever direction they'd take and departed, feeling sympathy for the weight of their judgment and grateful that it wasn't mine.

Déjà vu All Over Again
By Mario Fortin
Montreal, Quebec, Canada

I have driven my '41 Dodge Special Deluxe sedan street rod to a great many car shows all over Canada and the United States. My car is not the prettiest or most popular but I have modified it a bit to the way I like it and it drives pretty well with its 1973 Pontiac 350 engine.

I estimate I have driven the car 300,000 miles in the 32 years I have had it. With these many miles on the road I have had ample time and opportunity to have a few big accidents as well as "fender benders."

I left the 1998 NSRA Nationals in Louisville and was on my way to Cincinnati when I hit a particularly bad area of road construction with lots of dust, potholes and "Jersey Barriers." I looked to my left to check the traffic coming up on me and when I turned back around, I was right on top of a Jeep. I ran into his spare tire which pretty much tore up the headlights and my greatly modified grille.

A couple of the alterations of the car that got torn up pretty badly were the headlight bezels which were from a '76 Buick Skylark and some aluminum strips that I used for the grille. Didn't hurt the other guy's spare tire much.

I was able to drive the car so went back home to Montreal.

I repaired all the front-end damage at home in the next two months and headed out to the NSRA Nationals North in Kalamazoo, Michigan in September.

We've all seen the strips of truck recaps here and there along the side of the road. One of these rubber strips hit me after it came off a big tractor trailer that was ahead of me. It really did a job on the front of my car...again. I limped off an exit of the Interstate and found a radiator shop. A quick half hour repair at a cost of $20 stopped the leaking and got me on the road again. Back home in Montreal I changed the grille again, fixed the damaged headlight and put in a new radiator.

Forward to 2004. Passing through Las Vegas on my trip to Bakersfield, CA to attend the NSRA Western Nats, the choke on my carburetor apparently was closed too much and couldn't be fixed easily. It needed a new carburetor anyway so I picked up a new one in Dallas, Texas.

On my way to the Northeast Street Rod Nationals I shredded a fan belt. I made use of the Fellow Pages by calling Bernie West of the Champlain Valley Street Rodders who brought a half dozen of his old belts to me and one of them fit pretty well.

I lost third gear of my automatic transmission one time when I was about an hour out of Louisville on my way to attend the Nationals. I managed to get the car towed to a place where a fellow with lots of spare transmissions came up with a governor which cured the situation.

I plan to go to many more car shows across the U.S and Canada so I am hoping that there will be no more of these "mini catastrophes" because I have been there and done that and had enough!

Holy Piston
By Dennis O'Brien
Charleton, Massachusetts

My wife and I were returning home from the National Street Rod Association Nationals in Columbus, Ohio in the summer of 1987. My '38 Ford panel delivery truck street rod was powered by a 383 cubic-inch Chrysler engine at the time. I noticed more smoke than usual coming from the tail pipe and the engine was overheating and using oil at a rapid rate.

We were passing through Youngstown, Ohio at that time and we stopped along-side a pretty lake to survey the situation. It seemed like a serious predicament that couldn't be solved with a quick fix right then and there, so I got out my Fellow Pages and found a listing for Dick Nard who lived a mere five miles away.

I called Dick and he guided us to his home. He was a local disc jockey but also a very active street rodder who had built several street rods, so he agreed to help us out by providing shop space for work on the truck. We did a compression test and determined that one cylinder had very low compression. We took off the cylinder heads and discovered in short order that there was a hole burned in a piston, obviously caused by the overheating engine which led to pre-ignition and/or detonation. This was a big problem as it was, but the pistons in the engine were thirty thousandths oversize. That made it very difficult to locate a duplicate piston.

We went to a nearby auto parts store where the salesman located a place that had the piston we needed. He said it would have to be sent to us. I agreed to have it sent but later discovered that the place was only within a couple of hours drive so we could have driven there in Dick's car and saved lots of time. Bummer!

My wife and I stayed at a nearby Knight's Inn for a week to wait for the delivery. Dick and I got the engine all prepped for the new piston but still had lots of time to wait. My wife was so bored in that purple walled motel room that she said she never wanted to see the inside of a Knight's Inn ever again! Wearied as she was, it was not as bad as it could have been because a street rod friend, Steve Protchaine, and his wife and baby son stayed with us. We had come across them on our way to the Nationals. His '32 Ford roadster was parked along the road and Steve was inspecting the top. It had somehow split and had been "flapping in the breeze." We had some "racing tape" with us so helped him fix the top. They accompanied us on the trip so it was good to have them stay with us. One good deed deserved another!

We appreciated the Fellow Pages and equally valued the fact that Dick was so willing to have us use his shop for the teardown and rebuild of the engine.

We got back on the road again and finished the long trip home without any further problems.

"Thank you, Fellow Pages!"

Pumps in a Barrel
By Art Stultz
Colchester, VT

In 1983 my wife and I drove in our '32 Ford coupe street rod from Vermont to Berea, Ohio, near Cleveland to attend the Hot Rod Magazine Supernationals. There we enjoyed the show as well as a Cleveland Indians major league baseball game.

On the way home on the interstate, about 30 miles east of Cleveland, I could see a few drops of some type of fluid accumulating on the windshield. It was surely not rain on this sunny day. In a few seconds a torrent of fluid hit the windshield and the whole left side of the car. Radiator coolant? Engine oil? I pulled over and hopped out of the car to examine the fluid and where it was coming from. It was from the power steering pump which had partially come apart and was obviously not repairable. Various ways of proceeding ran through my head for the next few minutes. I had a roll of paper towels with me so was busy cleaning up the mess when a car pulled up behind me. It was not a street rod or custom car, but the two guys that got out of the car had car-themed tee shirts on so I knew I was in good hands.

I pointed out the mess and explained the power steering pump was done for. One of the guys said that if I could follow him over to his shop, he had a barrel of power steering pumps and I could pick one out! I simply took the power steering belt off and used "Armstrong Steering" to get to his shop, which was perhaps five miles away. What luck! He indeed had a barrel of pumps, maybe 15 of them. I pulled them all out and found one that was pretty close to my GM pump regarding the orientation of the inlet and outlet. He asked $10 for the pump. I put the pump on. He had some fluid, too, so I put that in and away we went.

The rest of the trip home took us through the Adirondack Mountains of New York State. While winding our way up and down through the hills and mountains, I felt the brake pedal was closer to the floor than it should be when braking. I found a wide turnoff rest area and inspected the brakes. Since I had a bottle jack, some wood blocks and basic tools I could take off the wheels, and shortly came across the problem at the right rear brake area. I found that a part had come loose but thankfully hadn't fallen out along the highway. This was so long ago I can't recall exactly what part it was. It might have been one of two struts between a wheel cylinder piston and the brake shoe. In any case, the displacement of the part allowed the brake cylinder pistons to come out farther than they should and thus the pedal to go closer to the floor to activate the shoes against the drums. I put it all together and was good to go. I didn't lose any brake fluid and the pedal was at normal height and firm as it should be. On the road again with no trouble the rest of the way.

This is a good lesson on bringing repair tools and equipment with you that you think you might need in case of trouble. Also, when traveling with other street rodders as they might need to borrow your tools as well.

Minor Details
By Frank Hanley
Leesburg, Florida

The late Bob Landry was my best friend all through Burlington High School (Vermont) back in the late 1950's and early 60's. We "hung out" together, especially after school, working at his place or mine on all kinds of projects, especially control line model airplanes which were the precursors to the radio-controlled models of today. Since we were both interested in mechanical things, we naturally graduated to taking an interest in cars.

In 1960 Bob came across a 1956 Morris Minor convertible which had been used and abused. He bought it from the Dodge dealership in town which was just across the road from where he lived and where his father David worked. The dealership was also a British Leland dealer. The engine smoked and the transmission was obviously "on its last legs." The first thing we did was remove the engine from the car by way of a chain attached to the cylinder head and a 2 by 4 run through the loop of the chain. With Bob on one side and me on the other we lifted the little four-cylinder engine out of the car and took it down to Bob's basement. We proceeded to rebuild the engine with new rings and bearings and a little valve work. The engine went in the same way it came out; with the chain and the 2 by 4. We had done our best but because of our young age and lack of experience we did just the minimum amount of work the best we knew how. Despite our efforts the engine still didn't run very well and the transmission, since we hadn't touched it, was about to come apart with any amount of heavy use.

Since Bob's father was the truck mechanic at the Dodge/Leland dealership he had a good connection with the owner and the mechanics, so it was easy to locate a worn-out high mileage M.G. engine and transmission that was stored in the back of the shop to be used for parts. After a talk with the owner of the engine and a little sweet talk from Bob's Dad, Bob was the proud owner of a beat-up engine and a good transmission. The original Morris Minor engine and transmission was pulled out again and retired from further duty. The upcoming engine change was not exactly the common small block Chevy in a hot rod Ford of today but sometimes you make do with what you have!

The M.G. engine needed a complete rebuild so the cylinders were bored out into the water jacket and then the block was sleeved. New oversized pistons and rings were installed and the crankshaft was polished. The head was planed and the valves and valve seats were reconditioned as well. A twin Stromberg carburetor setup was installed before the engine was put into the car. The compression ratio was about 10.5 to 1. With the larger displacement and twin carburetors, we "guesstimated" the horse power to be about 100. I think we were pretty close.

Even at his young age Bob was able to complete the swap by making whatever was needed to get the job done. To the best of my recollection parts were made in the machine shop classes Bob was taking at the Burlington High Trade School. A couple of the items that he made were motor mounts and the transmission perches. He also modified the drive shaft to fit.

Finally, everything was in, all hooked up, tuned and ready to go. A couple of mild test runs at moderate speed around the block, a few more adjustments and she was ready for the open road.

Route 2A in the nearby town of Williston was selected as our next big test run. I excitedly accompanied him in the "shotgun" seat. A short distance from the busy Taft's Corners intersection heading south Bob "dropped the hammer" and we proceeded to accelerate at a high rate. The speedometer went only to 80 miles per hour. We had long passed that speed when we came upon a Cadillac traveling in the same direction. We were by him in a flash and were feeling quite smug about the whole thing when the oil line to the oil pressure gauge decided to rupture at the point where it passed through the firewall. I was drenched with hot oil! The original engine produced only about 35 pounds of oil pressure but with the high rpm that this new engine was turning, the oil gauge was pinned at 65 psi. The oil soaked both my regular shirt and tee shirt and made it to my pants as well. We pulled off to the side of the road where I took off my shirts and wiped the oil off my face, main shirt and pants. I still carry a scar from the second-degree burns.

Meanwhile Bob lifted the hood to check the extent of the oil spray in the engine compartment and to see just how the oil line had come apart and what we should do to fix it. A few minutes later the Caddy caught up to us and stopped. The guy got out of his car, walked over and took a look into the engine compartment and asked "what the hell have you got in there for a motor?" "Just a little four-cylinder" was Bob's reply. Of course, we didn't tell him the horsepower had doubled from the 50 horsepower of the original Morris Minor engine.

Although I was hurting from the oil bath I took, we still had to get the car home. We folded over the oil line at the point where it was fractured and took two rocks and pounded the fold flat. Of course, no oil could get to the gauge but it allowed the engine to run and thus we got back to Bob's home.

Later Bob put a straight pipe exhaust on it and you could hear that Morris Minor coming a half a mile away.

Bob held his interest in cars throughout the rest of his life, right up until he died of pancreatic cancer at the age of 75. He had a nice shop at home that included a lathe and milling machine where he built street rods and tinkered with all sorts of mechanical things for himself and his friends. His pink and white 1928 Ford Model A panel delivery street rod that he called "Pretty 'N Pink" was well known in the greater Burlington area at all the local car shows where with its pink theme was the favorite with the young ladies. He never completed his Factory Five '33 Ford replica street rod, but I am sure that had he lived to finish it, it surely would have gone way faster than the little street rod Morris Minor with the hopped up M.G. engine.

A '56 Morris Minor convertible

Bob and Janet Landry with "Pretty 'N Pink"

The Obstacle Course
By Don Amundson
Auburn, Washington

In 2012 a group of us street rodders was coming home from the NSRA Golden State Nationals in Sacramento, California. We were about twenty miles south of Mount Shasta on Interstate 5, northbound. The guys I travel with all have CB radios to communicate with each other. This is a good practice because if you can pick up on information regarding highway driving conditions, incidents and accidents, or if someone needs to stop for a rest or get gas this helps a great deal. It is good for coping with road hazards and unexpected occurrences as well.

We were all going up a gradual grade rolling along about 65 miles per hour and approaching a long curve. The highway at this particular section had three lanes, the HOV (high occupancy vehicles) lane on the left, the center lane and the curb or slow-moving vehicles lane on the right. The guy in front of me yelled into his mic that everyone should move left to the HOV lane or to the curb lane immediately. We did indeed move to the left. Just around the bend was an elk carcass lying crossways on the line between the middle and the curb lane. About a quarter mile further there were three street rods sitting on the shoulder. The drivers and passengers were looking underneath their cars and checking the passenger sides of their rods as well. Two of the rods were badly damaged from running over parts of the elk. It could have easily been us! Farther up the road about a quarter mile was another rod on the shoulder. The owner was on his hands and knees looking under his car. It looked like the right-side muffler was hanging down four or five inches lower than the left side. Apparently, he had hit the carcass as well. Fortunately, this happened in the daylight. Had it been in the dark it could have been disastrous for all of us. We were really glad that we were on our CB's and had avoided what could have been a horrible incident.

Street Rod Farmer
By Norm Leduc
South Burlington, Vermont

As a kid growing up on my parents' Vermont farm in the 1940s and '50s I did necessary chores but also repaired many pieces of farm equipment. My three brothers and one sister and I worked hard when not in school as is the case with most farm kids. I never considered myself to be a mechanic but with natural curiosity and decent skills I did a pretty good job with my assignments. I became a pretty good welder as there was great need for this expertise on the farm.

My interest in farm machinery extended to automobiles as well. Someone, I can't remember who, taught me how to "hot wire" cars. Most auto ignition systems were easy to hot wire in those days as they all had a coil and "points" in a distributor and you could simply go to the switch up under the dash and connect "hot" to the wire leading to the coil. There were several easy ways to activate the cranking system as well. Don Ayers, who worked on the farm with me, lost his car keys so I hot-wired his '37 Oldsmobile for him. We were working with a ton and a half Ford truck one day and didn't have a key for it so with a simple hot-wiring job we got it started and drove it to the big town of Burlington, some ten miles away. Church Street was the main shopping venue in the downtown area then as now, so it was a big deal in those days for the young men to drive up and down Church Street with their hot rods and customs. This truck was certainly not a hot rod but we had fun with it anyway.

In 1955 when I was 16, I took a liking to a 1949 Ford two-door sedan and bought it for $100. It had a flathead six-cylinder engine which I learned later on was quite a rarity in that most of the Fords of that era had the much more popular V8 engine. The flathead six was built from 1941 to '51 and after that sixes were built with overhead valves.

I drove the car a while and decided that I should rebuild the engine. I knew next to nothing about car engine rebuilding but being curious (and a bit naïve) plunged into the job "with both feet." I totally disassembled the engine, replaced old parts with new and put it back together. Uh oh! Now it wouldn't start! I figured it was just too tight with all the new parts and wouldn't crank fast enough. My brother Maurice and I applied the old trick of chaining the car to our early '50's Farmall A tractor and pulling it around the yard. Maurice drove the tractor and I was at the wheel of the '49. After getting it up to speed I let out the clutch with the transmission in gear and after a half minute or so the engine started.

I drove it for a month or so but one day when I was a few miles away from home the engine blew! We towed it back home. Looking back at the "rebuilding" job many years later I realized I had not known enough to label anything like connecting rod bearing caps, lifters and valves as to their position in the engine so may have put pistons in backwards or switched connecting rod caps around and put bearings in incorrectly or in a sloppy manner. I was discouraged with my second-rate ability as an engine builder so bought a version of the more popular Ford V8 engine from a friend and put it in. I didn't have much trouble simply switching one engine for another and didn't dare to rebuild it so that went fine. I had to get another radiator as the one in the V8 had two coolant inlets and two outlets whereas the six had just one of each. The loss of the flathead six may have been a blessing in disguise as the V8 had so much more power and speed than the six. I drove it around the county and into town often getting a bit heavy on the gas pedal for the next five years. I lost track of the car after I sold it so perhaps it is still in the hands of a "gearhead" car nut somewhere. Maybe even a street rodder!

A Busted Generator
By Doug Hughes
Sarasota, Florida

My girlfriend Marie and I go up to Colchester, Vermont from our home in Florida most every September to visit relatives. We get out Marie's stored '57 Chevy and go to the annual Lake George, New York Car Show. On the weekend of September 5-8, 2013, we were enjoying the show as usual. One of the main features of the show is some impromptu "cruising" on the boulevard and the neighboring streets. The Belair was going along the boulevard just fine when suddenly there was a racket under the hood and the "tell-tale" charging light came on. Uh-oh! Generator trouble!

The engine was pretty much stock, but was modified somewhat: The original generator was mounted to a later model exhaust manifold which had a mount for the generator bracket.

I'm not a real good mechanic or street rodder, mind you, but since Marie loves her '57 and we get lots of enjoyment out of the car, I tackle most of the basic maintenance. I know just enough about cars to do a few repairs on it.

Upon opening the hood, I could see the nose of the generator was headed on a bias and the belt had flipped off its pulleys. The front bearing of the generator had disintegrated and the front of the armature shaft had actually broken off. Hmmm...I'll just go to the nearest auto parts store and buy a new armature for the generator. Little did I know that you can't get a generator at an auto parts store let alone an armature for one. The parts guy suggested upgrading to a "one wire" alternator, which is popular with hot rodders. I did so. Next was to find a way to mount it. I had met a local guy who had some odds and ends in his garage nearby, which included an old bed frame. The side rails were essentially angle iron, so I used his hacksaw and cut off a four-inch-long piece of it. He had an electric drill but we had to go to a hardware store for a drill bit that was the right size, and an array of bolts, washers and nuts. I drilled holes in the ends of the angle iron and then mounted one end of it to the alternator mounting tab and the other end to the original generator bracket. This gave the alternator a good position, although it was a bit shaky. The belt was a bit short, which put the alternator on the rocker arm cover. I didn't want to go back to an auto parts store for another belt, so I put a foam pad between the cover and the alternator to act as a cushion.

Now...what about the wiring? I had a general idea of the generator wiring and figured that if the generator put current out from the "A" post, then the wire fastened to it should be fastened to the alternator output post. This seemed to work as the battery didn't run down when we drove back to Colchester. The tell-tale light stayed on however.

I had a Colchester street rodder, Art Stultz, work on the car a year before so we took it to him a week after this debacle and he changed the wiring to make it function properly, including eliminating the tell-tale light. When time permits, I will find a proper alternator mount from a parts catalogue and make the mounting bracketry correct so it is more stable. With a belt the proper length I should be able to adjust the belt tension. Our beautiful Belair will be just fine!

Bikes Have Floorboards Too!
By Brian Martin
Underhill, Vermont

I have owned several street rods over the years but as with most all rodders, we have other interests as well, such as motorcycles.

My bike of choice back in 2007 was a 2005 Harley "Dresser." On October 2, (I remember this date very well!), my friend Gary Blodgett and I decided to go out cruisin' on our bikes. He lived in Huntington so I met him at his home to go to the town of Warren for lunch.

Gary started off ahead of me on his extremely modified 1976 "Shovelhead" Harley and I followed. Gary likes to "get on it" much more than I do so he was way ahead of me when I approached a curve.

The previous owner of my bike was pretty short so he had installed a three-inch lowering kit to make it more comfortable for him. The "floorboards" of this modified setup were thus pretty close to the road so when I hit the "negative cambered" curve the floorboard on the side I leaned toward hit the pavement and I went flying! As I headed off the road and toward the woods of this remote area, I had to think quickly about whether it would be best to stay on the bike or push it away. I timed it so I could get to the best area from which to push the bike away from me and land the best I could. I first landed hard on the edge of the road breaking my "tailbone", then came to a sitting position, kicking up stones which covered my face and gloved hands. I then went airborne. I couldn't see the tree coming but it surely came! I plowed into the tree branches breaking several of them with my arms. I continued over the steep bank maybe twenty feet or so below the roadway where I came to rest.

Meanwhile, Gary realized I was not coming along at the rate I should have so turned around and headed back. When he found me, he took off his tee shirt to use as a tourniquet to stop the bleeding. I told Gary that we had passed a house not too far back where a man was mowing his lawn so he should go back there to get help. He did so and the fellow called 9-11 then grabbed a blanket and drove in his car behind Gary to the scene. In a short time, several fire and rescue trucks from the nearby town of Richmond arrived. They used chainsaws to clear a path down the bank to get to me. I am a big guy who weighs 240 pounds so it took quite a bit of effort to get me up the bank, into the ambulance and off to the big hospital in Burlington.

A long involved twenty-hour operation ensued in the attempt to put my leg back together. Both the big bone (tibia) and the smaller (fibula) were broken with the distal ends hanging loosely with about two inches of the tibia completely missing. They inserted a long pin through the upper and lower remaining parts of the tibia to re-establish continuity. They took a piece of thigh skin to fill the space and promote bone re-growth. There was some vascular "mini-surgery" as well.

I spent six weeks total in the hospital, which included the last two weeks in physical therapy.

After a year, I was able to go back to work as a self-employed fine interior woodworker. I get around okay but I limp a bit. I am able to do my work pretty well. The bike I was riding was totaled so I sold it for parts. I sold my three other Harleys as well. I have decided that I would be better off continuing with the build-up of my recently acquired '55 Chevy station wagon. The floorboards of this machine are more to my liking for sure!

Wiped Out on the Big Trip
by Jim Higgins
Weymouth, Massachusetts

In 2014, my friend John "Ducky" Duckworth of Maine created an excellent "road tour cruise" that took our several street rods and customs throughout the East Coast including Virginia and featured attending the big 2014 NSRA Nationals in Louisville, Kentucky. As it turned out, it rained almost from start to finish of the first part of the trip.

The driver's side windshield wiper on my '50 Mercury was adjustable but wasn't set quite right. It kept hitting the windshield's rubber gasket. Well into our trip the rubber of the blade tore right off.

We pulled off the highway at a rest area and thought it was a simple, straightforward idea to transfer the passenger side wiper over to the driver's side. Not quite so easy, however. A tab on the right wiper did not line up correctly with the notch on the left side, so it simply did not work. Not too long after we had gotten back on the road, the wiper blade came off and went flying by the car on the left side. That was the end of that idea!

With the continuing rainfall and no good idea of trying to find a wiper arm that would fit, we drove on to Louisville with the tactic that I would drive right behind Ducky in his '55 Chevy station wagon. Ducky's car had nice LED taillights that he kept on to make it easier for me to see despite the deluge of rain. We made it okay, although it was not fun for sure!

In Louisville, I picked up a wiper of the correct design so was all set for the rest of the road tour.

Oh Deer!
By Stan Morrow
North Ferrisburgh, Vermont

I purchased a 1930 Ford Model A coupe in 1978 from a friend who lived nearby. He had gotten it from a fellow in Ellenburg, New York. I worked for two years making it into a street rod. It featured lots of changes including installation of a 1980 Chevy Monte Carlo V6 engine. I painted the car yellow.

It served me well for thirty-five years as I went to a great many car shows and on lots of cruises in New England, Ohio, Canada and elsewhere. The little V6 did remarkably well providing reliable power for many years.

I am one of the founding members of the Champlain Valley Street Rodders so I drove the car to club meetings quite often.

Street Rod Horror Stories

In June of 2015, my neighbor and fellow club member Chet and I went to a meeting in St. Albans, Vermont in the '30. After the meeting, about 9 PM, we drove onto Interstate 89, a major interstate in northern Vermont. We headed south going 60 to 70 miles per hour past several exits and suddenly from the left side a big deer bounded at us from the median!

It cleared the left front of the car but hit about midway in front striking the grille and right front fender. His body then swung off to our right and to the side of the road. I hit the brakes of course, and headed for the breakdown lane as I was broken down without a doubt! I limped ahead for a short distance looking for a good place to stop. Almost immediately a fellow pulled over to help out. Neither Chet nor I were hurt but the car was pretty much done in with lots of front-end damage. We borrowed the guy's cell phone and called 911 as well as Chet's wife, Connie.

A State Police officer and the Colchester Fire Department arrived shortly. My flashlight had dead batteries but the officer's flashlight did a great job of lighting up the car and the area to reveal that the engine's fuel filter was leaking. Being equipped with my tool box and assorted fasteners and things I found a bolt the right size and stuck it in the fuel line rubber hose and stopped the leak. There were no other problems that needed immediate attention. The fire department guys put out cones and milled around for a while...then picked up and left. A flatbed arrived and loaded up the car and took it to my home. Connie came and picked us up and dropped me off at home as well.

It has been over three years now that I have been without my dependable Model A but the insurance company has served me well so I have been able to purchase some aftermarket parts such as both front fenders, grille shell, hood, radiator and bumper.

While waiting for the new parts to come and deciding when to rebuild the car, I built up a '47 Ford pickup as a street rod. I am enjoying driving it around, including to club meetings. I keep a wary eye out for deer as I drive on the Interstate!

I'm showing off my '30 to several of the younger generation.

Street Rod Wedding
Mike Burnham
Milton, Vermont

Marriage is a very special event for everyone. Some couples may want a traditional wedding while others might prefer to make it unique by planning a wedding with a theme that is special to them. SCUBA divers might want to get married underwater with themselves and the wedding party in breathing apparatus and decked out in their gowns and suits. A couple that met at a riding academy might choose to get married on horseback.

Prior to my wedding I had finished my 1930 Model A Ford street rod except for final paint. My "intended" agreed that in keeping with the era of the car a "gangster theme" might be fun. You know, Bonnie and Clyde, Baby Faced Nelson, machine guns...

In addition to the "A," I owned a 1939 Buick that had been in storage since I was ten years old. It hadn't been running for a number of years but I got it going for the wedding, so we had both the A and the Buick to help provide the gangster theme. I recruited my brother to drive it.

On the day of the wedding, there was a big car show not too far from its site. Some of the guys in the wedding party, including myself, wanted to attend the show so we devised a plan to leave the show in plenty of time to get to the 4 PM wedding. I did just that and went on home in the A and waited for my brother who had forgotten his tuxedo. He had not arrived by 3:15. This was in the era before cell phones so I could not easily contact him to find out what was going on.

At 3:30 I gave up on him and headed to the wedding. On the way my brother went by me going in the other direction. His tux didn't fit so he was going back to the tuxedo rental place for a refit.

I was pretty anxious at this point worrying about my brother and the time slipping away and wouldn't you know it... the Model A started jumping and bucking and obviously failing for some reason. I pulled over and tested the spark plug wires while dressed in my tuxedo! I finally figured I had spark all right so perhaps I was out of gas. My brother came by at that time so he went to get fuel. He came up with a junk gallon container for the fuel but no funnel which was pretty much necessary for the fuel transfer with this particular car. We had a roll of paper towels with us so we unrolled the towels and used the cardboard center tube as a funnel.

All the time I am thinking that the bride is the one that should be late, not the groom! I finally arrived 45 minutes late! Everyone was out in front of the church waiting for me.

After the wedding, there was a picture taking session and then attendees left for the reception ten miles away. The time that it took for these events was enough for some of the guys to go out and pick up a half dozen new small gas cans which they lined up on a table at the reception. Quite a surprise and quite a joke on me!

The priest who performed the ceremony of course congratulated my bride on being on time and humorously chided me about being late.

Oh well, better late than never!

Picture used by permission of the artist, George Trosley

Rumble on the Highway
By Stephanie Paris
Ipswich, Massachusetts

One morning in 1980, my husband, Lionel (whom we all call "Puddy"), our two kids and I left for the NSRA Street Rod Nationals in Memphis, Tennessee. We were in our 1933 Chevy coupe street rod, which was pulling a homemade wooden trailer we had borrowed from the New Hampshire Street Rod Association. We took advantage of the space in the trailer and loaded in more stuff than we normally would take on a trip like this. Puddy was driving, I was "riding shotgun" and the kids were in the rumble seat. We had removed the partition between the rumble seat and the bucket seats that Puddy and I occupied so if it rained the kids could "hunker down" and keep dry. We made our daughter, 16, and son, 14, wear motorcycle helmets while we were traveling, which they were not happy about. I told them if we ever got into an accident they would probably go forward and hit their heads on the metal structure surrounding the rumble seat, so the helmets were for safety purposes. They understood that once we were off the highway they could remove them, which placated them somewhat. We made it to Virginia just fine the first day and stayed overnight in a motel.

We were back on the road the next morning, but not for long. Puddy was driving his usual high speed in the passing lane when we saw a solitary wheel and tire pass us and veer off the road. The kids immediately started hollering. I looked through the back window of the car and saw the trailer still hitched to the car swing around towards the driver's side of the car, then saw the other wheel and tire go by. The trailer without its wheels was now sliding terrifyingly on its axle with sparks flying. Puddy managed to pull the car to the side of the road. What should we do next?

Just then, several other street rodders came up behind us and stopped to help. They just happened to be following us, probably headed for the same show. A fellow in a big early '70s Chrysler Newport station wagon also stopped. This very nice man said he would be willing to transport the trailer to get it repaired. The street rodders took everything out of the trailer and put it into the back of the station wagon. Then they turned the trailer over and put it on top of the luggage rack on the wagon and tied it down. We looked around for the wheels but they were nowhere to be found. They were beat up and rusty and apparently some violent lateral shifting of the trailer caused them to pull off over the four lug nuts, and they went rolling off the highway out of sight.

The man said he knew of a place in Roanoke that serviced horse trailers and thought they probably could fix our trailer. We followed him there and unloaded all of our things. The trailer place fixed the bent axle, replaced a broken spring and put on new wheels and tires. Puddy wanted to give the man with the wagon something for helping us get to the trailer place but he wouldn't take it. He said he was just happy he could help us out.

Our two teenagers were pretty good about the long wait, although they were somewhat bored. We complimented them on being patient. After spending most of the day having the repair done, we stayed overnight in a local motel as it was pretty late in the day to resume the trip. We had allowed extra travel time for the trip so didn't miss any of the show.

Someone who heard about the incident notified the New Hampshire Street Rod Association, and they called us when we got home and told us to keep the trailer.

This accident made us very aware of the need to prepare both our street rod and trailer before going on a long trip. They need to be checked for safety and should be road tested to see if the combination handles okay. We can't always trust a borrowed trailer as you don't know just how it was built and maintained before the loan.

Oh Canada! Trouble x 8
By Art Stultz
Colchester, Vermont

Several guys in our street rod club had heard of the big Atlantic Nationals car show held in Moncton, New Brunswick, Canada each year and would discuss going to it once in a while. We finally decided in 2006 to get a group together and go up to it from our area in Northwestern Vermont.

Four of us met on a Wednesday morning in July and headed east. Stan in his '30 Model A which was pulling a utility trailer that he had made, Bob and Janet in their '28 Model A panel delivery, Harold in his '41 Chevy pickup and my wife Marion, and me in our '55 Chevy.

Things went well for a while as we enjoyed cruising the pleasant back roads through eastern Vermont and into picturesque New Hampshire. It was finally time for a pit stop and coffee. Stan reported that the trailer was really acting up behind his A with some obvious vibration and bouncing around, which seemed to be indicating wheels very much out of balance. We had reservations to stay the night at a certain motel in Bangor, Maine, that night so perhaps he could tough it out and put up with it until we got there.

We got to the motel in the late afternoon and Stan located a repair shop near the motel that could do the balancing first thing in the morning. Surely that would take care of his problem.

I asked the clerk at the motel desk if there was a good place to eat nearby. We were all tired and hungry and deserved to wind down for the day. He gave me directions and we all headed out with Stan riding with Harold. As we approached the classic stainless-steel diner, we could see lots of street rods and customs parked in the parking lot! What luck! There was a Wednesday Night Cruise-In right there where we planned to eat! A couple of guys directing traffic obviously thought we were there specifically for the cruise-in and directed us to parking spots. Sure, why not! We opened our hoods and the locals who had never seen our cars flocked over to see what these out-of-towners had to offer.

The next morning Stan got his wheels/tires balanced and we were off, headed for the Canadian border. Not so fast. We were cruising along when someone realized that Stan was not with us so we all turned around and found him a couple miles back once again examining his trailer. It was still acting up. The shimmying and vibration were even worse than before! Stan was now thinking that it must be something other than wheel balance such as a severe case of toe-in, as he had done extensive work on the axle and spindles so maybe didn't get the spindles squared up with the frame of the trailer as they should have been. We all discussed what to do. Stan was so disgusted he was literally ready to pull the trailer up into the nearby woods and either abandon it or retrieve it when he came back from Moncton a few days later. Leveler heads prevailed so we made the decision to take the wheels off to make it narrower and lift it into the back of Harold's truck. It just barely fit so off we went.

We crossed the border into Canada. About an hour later traffic dramatically slowed down. There was construction going on ahead but there weren't any warning signs. There was a chorus of screeching brakes from all of us and several other cars but we stopped nose to tail without any contact. That could have been a major collision indeed! Problem averted!

We arrived in Moncton at the motel where we were to stay several nights. Stan and others unloaded his trailer from Harold's truck and left it in the motel parking lot for the duration. Marion and I spent some time cleaning up the car, me on the outside and her detailing the carpets, seats, dash, etc. on the inside.

The next morning my car doors wouldn't open. Operating the transmitter for the handle-less doors did nothing. No action. Perhaps it was the transmitter battery but possibly the car battery was low on charge. Harold gave me a boost with his truck and battery cables. I had installed a pair of posts along the frame of the car just for this purpose. Then I could open the doors and start the engine but with a bit of difficulty. Off we went to the show which was some eight miles away. I found a shady spot to park that would be good for long-term parking as well as to diagnose my problem. The cranking motor would not spin over, just the common grunt. I had my tools, including a 12-volt check light and multi-meter plus my notebook of wiring diagrams. I always make up diagrams when I am doing wiring projects on my cars and have used them on several occasions. There seemed to be a fairly serious current draw somewhere that was drawing the battery down. With the help of a nearby street rodder we got closer and closer to the current draw and then... there it was! The driver's power seat switch was on! The switch would spring back to neutral after putting the seat forward but would not return back to neutral after putting it all the way rearward. I had been aware of the bad switch for some time and was used to putting it in neutral each time I had made an adjustment. However, I had failed to tell my wife about it! Apparently, the evening before she had moved the seat back and forth to clean the carpets and the switch had been left in the defective non-return position. With it in that position I could hear the seat motor attempting over and over to move the seat rearward and not able to as it was at its limit already. I moved the switch to the off position and that took care of the problem. But not entirely. The battery was old anyway and with the seat motor trying all night to operate the seat, it took the battery way down and it was not able to crank the starting motor. Fortunately, there were vendors at this show including a NAPA dealer, and as luck would have it, he had a battery on display just the right size as mine so I bought it and put it in the car. I also purchased a "battery pack" or "jumper box" which was gaining popularity at that time just in case of further trouble.

The show went on for several days, and then the other guys in my group parted for home and my wife and I headed to Nova Scotia to visit my two cousins. While there for several days I noticed that when turning hard left or right the brake pedal would go down quickly and I would have to pump a couple of times to build the pressure back up. Hmmm...Rather disconcerting...Was this worth worrying about?

It is hot during the summer even in Canada so used the A/C, but noticed it was not up to par and not keeping us quite cool enough. I found a shop that did A/C work but when the mechanic and I looked it over we determined that there was a leak somewhere, probably in the condenser, and I thought it was not worth pursuing as I didn't have much time. Rest of the trip: no A/C. Not the end of the world!

When we left for Vermont it was raining like mad. We had reservations for the huge "Blue Nose Ferry" that would take us from Nova Scotia to Maine, so couldn't wait any time for the rain to cease. It seemed like forever that we drove through the rain down the length of Nova Scotia, with the heating and A/C system not clearing the inside of the windshield very well and a water leak on the passenger's side sending a stream of water into the area at my wife's feet which she wiped up now and then.

I made a "tilt front end" to this '55 that in general works pretty well, with an electric motor (power seat base) that moves hood, fenders, grille, bumper, etc. forward seven inches and then I tip it up from there on hinges. The rough roads in Nova Scotia had caused the rear of the left front fender to come dislodged, so when I got to the ferry dock and about to pay the fare, I couldn't get out of the door without having the door crunch into the fender. I had to climb over the console and get out the other side and go around and fix it.

The next day or two went pretty well until somewhere in Maine I filled up the gas tank and, although I am usually careful to not fill it up full, this time I wasn't paying attention and filled it too much. The gas found a small leak somewhere and started dripping on the ground. I had deleted the gas fill door and made my own filler in the left tail light housing area like the '56's but didn't do a good job.

We hightailed it out of there to burn off the gas to get the level down so it wouldn't leak. We were ready for an ice cream break but although we passed a nice ice cream shop we didn't dare to stop just yet.

Back on home...

I later asked some guys in the club about my brake problem and two of them knew that if you put the brake caliper at the front of the rotor, then when an extreme right or left turn is made the caliper may hit on the suspension and push brake fluid back into the master cylinder and cause the problem. I eventually moved the calipers to the rear and took care of that.

I had a need to use the battery pack a month or so later and it wouldn't do a thing. Had to get a new battery for it. So, it wouldn't have helped had I needed it on the trip.

So: Trailer wheels. A near smash up. Defective seat switch. Worry about brakes. No A/C. Water leak. Displaced fender. Gas leak. Other than all that it was a pretty good trip!

Dad Was Mad - Story #1
Racing Rudy

Ken Bessette
Williston, Vermont

In high school, I wanted my buddy Rudy's '51 Chevy really bad because I thought it was a pretty cool car. I had visions in my mind about how I would fix it up if it were mine. All I had at the time was a pretty nondescript Nash Rambler but it could race around the streets of town pretty well nonetheless. Our school noon hour was quite long and we were allowed to leave the grounds. One noon hour I was going south on Willard Street in my Nash and Rudy passed me going north in his '51...my "dream car." I made a quick U turn to chase after him. I caught up to him just as he took a left turn onto Main Street. The traffic light was green for him but red for me---I went through it anyway and took the left turn practically on two wheels to catch up with him. The race was on!

Unbeknownst to me my Dad was stopped for the same red light coming toward me so saw me go through the red light much too fast for the situation. I didn't see him at all.

Rudy and I raced down Main Street hill for a quarter mile or so, turned left on Union Street and continued for quite a few blocks until we slowed at a rotary. Since Dad had a shorter route to drive than we did, here he was at the rotary when I showed up! He motioned for me to pull over, which I did. Boy was he mad! He admonished me severely for what he saw me doing and told me not to drive my car anymore!

Since I was there in my car in the middle of the day, he let me go back to school. I was taking Trade School Auto Mechanics and told Dad that I had arranged to have the class work on my car to fix some things that needed work. So, I left the car at the school that day. For many days after that, I took the bus to school and back, but was able to drive my car around town after school for an hour or two each day! Whoopee! That worked out pretty well! He never found out about that last part, much to my relief!

My '51 Chevy "Dream Car"

Dad Was Mad - Story #2
Over the Wall
Ken Bessette
Williston, Vermont

In my senior year of high school, 1956-57, my car was a '51 Chevy Deluxe two door sedan. It had belonged to my buddy, Rudy, but I had finally talked him into selling it to me. It had the basic six-cylinder engine, but I craved to change it to something with more pep and power. I concocted a plan that would please any street rodder with a mild six-banger in his ride. I would simply run it so hard I would tear it apart or burn it up. This would give me the necessary "excuse" to make a change. I accomplished the feat in short order so was looking for an engine that I could slip right in pretty easily but would have much more to offer on the streets on Saturday nights.

In those days, we didn't do a lot of switching big V8's where a six-cylinder resided as we were just learning street rodding as well as not having much money for the job. I went to Barcomb's Wrecking Yard and found a '53 Corvette engine (of all engines! What luck!) and made the simple switch. The engine had three single barrel carburetors and a split manifold with dual exhausts which gave me a pretty potent machine to tool around town in and look for an occasional drag race. I selected some mufflers with the well-known "deep mellow tone."

My Dad didn't like the loud exhausts and kept telling me to change the system to quiet it down. Of course, I resisted as long as I could. To keep him calmed down, I found a way to park the car at a distance from the house, so the sound of it wouldn't be so bothersome to him. This would allow me to postpone making the change to the exhaust. Our house was off of the main road just a bit. Cars coming to our house would go from the main road down the driveway, curve to the left and head for the garage. The lawn next to the driveway ended at the top of a stone retaining wall which was about seven feet high. There was a convenient place right up against the wall for Dad to park his '55 Chevy Bel Air.

Back at the main road there was a little "pull-off" area on the same side of the road as the house. I would come back home after being out on the town, put the transmission in neutral, shut off the engine and coast into the pull-off area and leave it there for the night. I would then simply walk down the driveway and into the house.

One morning a tractor trailer roared by my car in the pull-off area. I had forgotten to put the transmission back into gear from neutral the night before, so the car had just enough impetus to start moving off the road, down the slight grade and across the lawn. It rolled to the top of the retaining wall, partially over it and down on top of my Dad's beautiful Chevy! It landed over the driver's door area of the '55. There was quite a bit of damage to both cars. My car hung right there with the back of it still on the retain-

Street Rod Horror Stories

"Well Dad, I had to park it some place".

ing wall. We had to get it off with a wrecker truck. Hmmm...should have gotten those mufflers changed and parked in the garage to begin with! Dad was very angry to say the least!

Dad Was Mad - Story #3
Rocket Man
Ken Bessette
Williston, Vermont

When I was in high school in 1957, I had a "hopped up" '51 Chevy. Sure, I ran it pretty hard and was always breaking axles, tearing up the manual three speed transmission, etc. One Saturday night, I was anxious to go out somewhere but my car was not working for one reason or another.

The big Memorial Auditorium in my hometown of Burlington, Vermont was a very popular place for basketball games, shows and conventions. A block away was the YMCA. This particular night, the annual Firemen's Ball was at the auditorium and my Dad wanted to attend it. I arranged to ride in with him in his big '49 Oldsmobile Rocket 88 and then I would walk over to the YMCA and hang out with some of my friends while he attended the Ball. The auditorium was about midway down Main Street, a very busy street and a pretty steep hill. Dad found an angled parking place on the very end of all the parking spots. What luck...this was the best and nearest parking spot to the auditorium.

The three other guys who were at the "Y" and I became kind of bored with not much to do there and decided to cruise around town...in my Dad's Olds! After all, this thing could do 100 miles an hour easy! (So I told them!) Dad had the keys to the car of course and wouldn't approve of us taking the car anyway, so I used a "church key" (bottle opener) to "hot wire" the ignition switch. That model of car in those days had a simple on/off ignition switch where I could jam the church key between the two posts to get electricity to the coil. There was a separate stock momentary button switch for cranking. Good to go!

We cruised around town for a while and then went down to the south end of town where I knew of a pretty good out-of-the-way road where I could "open her up"! On a particularly little traveled part of the road, I floored it and got it up to near 100 MPH. We soon approached a curve, and the back end of the car came around to the right, set the rear wheels onto the gravel, and sent the car sliding backwards into the ditch. The church key came dislodged from the ignition switch and thus the engine shut off. When we came to a stop and things calmed down, we got it started again and got it out of the ditch and went back to town. No apparent damage to the car.

Uh oh! The parking spot the car had been in was taken, so we went up the hill a way and found an open spot. I knew Dad could easily remember that his car had been parked in that prime spot, so I wanted to get it back there before he came out of the auditorium. We waited there awhile until the Firemen's Ball was over. One of the first people out took his car from the spot we were waiting for. Ahhh...good! I used the church key once again but as we approached the spot we wanted to go to, it fell out of its precarious position once again. The engine quit and as I tried to maneuver the car, it jumped the curb so I just let it sit there.

In just a minute or less Dad came from the auditorium and approached the car. The three guys with me hightailed it out of there and I sheepishly said we were just sitting in the car to while away the hours. However, when Dad started up the engine, he could see the temperature gauge at the normal operating temperature, so he put it all together and figured out what we had done. He was pretty upset. I was in the doghouse for a while after that!

Had to Bag This Trip
By Bob Powell
Belleview, Florida

There are many styles of suspension on street rods and customs such as original leaf spring suspension, torsion bar suspension, coil spring suspension as well as a very sophisticated system which utilizes air bags. Since air compresses and decompresses it provides a pretty agreeable ride. There are several vendors across the country that sell air suspension and hydraulics.

I am a member of "The Losers" car club in Wisconsin. A fellow member in the club had a '39 Ford which had air suspension on all four corners of his street rod. Each corner of the car could be raised or lowered with individual controls.

When we were out on a cruise one day, the left front air bag failed which caused that corner of the car to come down with a resounding crash and the fender to slam hard onto the tire. The car immediately went into the left lane before the driver could regain control and bring it to a stop.

We discovered that during suspension travel over a period of months, the upper ball joint rubbed the airbag, causing the failure. Running that bag/shock assembly upside down would have prevented the interference. He converted the air suspension to "coil-overs" a short time later.

I have been an NSRA Safety advocate for years and it seems like there is a great need for a "fail-safe" system to be implemented for obvious safety reasons. With fendered cars it is a must to keep the tire from hitting the fender, severely damaging it and causing a bad accident. I know of a fellow who built a simple bracket that he has on his full-fendered '32 Ford that looks like it might work and prevent a "horror story". Unfortunately, car owners want to slam them to the ground while at events. I've always said, "If it looks right going down the road, it looks right in the show."

"Don't Be Judging People"
Jim Rowlett
Hurst, Texas

I have been the long-time chaplain of the National Street Rod Association. In 2013, I drove my '37 Ford street rod on the first NSRA Road Show Tour from the Rocky Mountain Nationals in Pueblo, Colorado to the Northwest Street Rod Nationals in Ridgefield, Washington. The tour ran from June 17 through July 6. There were street rods and customs in our group.

Since we were all, of course, avid "car guys" we looked over many of the other cars in the group now and then. One car, a '63 Buick Riviera had several "questionable parts" and equally dubious craftsmanship under the hood. Several of us were shaking our heads and asking ourselves if this car could indeed make it through the long drive without some breakdown. Among other things, the headlights and radiator hose clamps looked original and had probably seen many miles. This caught my attention and imagination as to what else unseen could invite tragic situations before long. Passing through Dumas, Texas one of the headlights on this car failed so was replaced and in Butte, Montana the other one went out and was replaced as well. In Coeur D'Alene, Idaho the engine was running hot and a new radiator cap seemed to take care of it. Surprisingly, the car continued the rest of the way on the tour without any problems.

I was feeling pretty smug with my assessment of this car and confident that my own car was surely in first rate shape. Before each trip I always inspect and fix anything that I see that needs repair. Surely this eternal vigilance would pay off handsomely. Nope, not the case.

My first problem was with the distributor. It took a little doing but it was repaired.

I had some trouble with the fan motor in Washington State. I worked on it and thought it was fixed.

In Pueblo, Colorado a fellow in our group who was following right behind me spotted a low tire on the luggage trailer I was pulling behind my car. I corrected the pressure at the next convenient stop and continued to drive. The tire held up for quite a long time but in Utah I got a cell phone call from the same fellow who was again behind me. He said the identical tire was going flat again. We stopped on the side of the road to change the tire and continue on our trip. At the next stop, as the group prepared for lunch, one of the guys took me and the leaking tire to a truck stop to see about repairs. I scratched my head a bit and did some thinking about how to test it. Even though we couldn't see any problems with a test with soapy water, the fellow at the truck stop simply bent the stem of the valve over and out swooshed some air! A new valve took care of it.

My next problem was with a power brake booster check valve. After I had taken part in an early Friday morning TV show on the fairgrounds, the car started idling poorly. Initially the distributor was diagnosed as the problem, but after it was replaced, we still had the same idling problem. The check valve fixed that issue and also corrected a problem we had with the brakes!

On Sunday, after leaving the fairgrounds I had trouble with the fan motor again. This led to a couple hours delay by the roadside until we could get going again.

I was critical about the other fellow's car and cocky about my own but that attitude came around to bite me as I ended up with more problems on that trip than he did! I have since spent time with the owner of the Riviera and explained my unfounded concern about the car. He told me that he was very unsure about it himself! We both had a good laugh about the situation!

So...a pretty good lesson for me: "Don't be so quick to judge other people or their cars!"

Real Flames
By Walt Kruger
Walnut Grove, Missouri

In 2011 my wife and I were on our way from our home to the 8th annual Fall Spooky Moon Hot Rod Picnic in Everton, Missouri. I was driving my 1936 Chevy two-door sedan street rod when I noticed smoke coming out from the sides of the hood. I pulled over to the edge of the road, shut off the engine and opened the hood. The Holley carburetor had flooded out so that fuel had overflowed and then it somehow caught fire. I thought of an old trick that many car guys know about: I quickly started the engine again and revved it up which sucked the flames into the intake manifold and put out the fire inside the carburetor. I also smothered the fire on the outside of the intake manifold area with a heavy towel I had in the car. I give full credit to the K&N flame proof air filter for helping to save the engine as well.

The rear float of the carburetor was too high so I readjusted it to the proper level. The engine ran very well but I had to do some major cleanup of the engine compartment when I got back home from the trip, as a result of the brief fire.

In 2014 I sold my '36 and bought a '56 Chevy Sedan Delivery. The car was built in the Los Angeles GM Assembly Plant as VIN #36. The engine is a '69 Camaro SS 350, the transmission is a 700R4 and the rear is a "12 bolt" from a '67 Nova.

My wife and I took the '56 on a cruise of about 150 miles from home to the Lucas Speedway in Wheaton, Missouri. When we completed the drive back home, I pulled back into my yard and heard a horrible metallic sound coming from the left front of the car. I jacked the car up and when the tire cleared the driveway the wheel was so loose it practically fell off! I found that the front left wheel's original ball bearings were partially disintegrated. The questionable thing was that I had serviced the brakes just a few months previously which included inspecting and greasing the wheel bearings.

Wow! What good luck! Good luck, if you look at the bright side, in that this potential disaster happened just as I pulled into my driveway after such a long trip.

After that scare I added new front disc brakes, front and rear springs, tie rods and shock absorbers and topped it off with new power steering as well. What a difference that made to the handling and braking!

On Memorial Day weekend of 2017 I attended the NSRA Mid-America Street Rod Nationals in Springfield, Missouri. On Saturday, the busiest day of the three-day show, weather reports were warning of impending high winds and damaging hail. I helped some of the other cars get under cover within the many fairground buildings that house animals and equipment during the annual fair. I then took off for home which was just 17 miles away. The skies were looking pretty ominous so I was looking forward to getting my car safely in my garage before any damage was done by rain and hail. Just two miles from home a doe came charging full speed from the side of the road and crashed into the driver's side front end. She smashed my headlight and damaged the grill and some chrome trim and although she probably had internal injuries, she got up and ran off. Not much damage to the car because it was a '56 and not a 2016! This didn't stop me from returning to the show the next day however, as the car was still drivable. I went off to the show minus one headlight and with a mess of whitetail hair! I got the damage fixed later to the tune of about $1200 and am good to go for another season!

Tragic Fraying
By Gene Tinney
Bourbonnais, Illinois

A 72-year-old friend of mine in my state of Illinois was a long-time street rodder who owned five street rods. It was pretty well known in the area that he drove his cars pretty fast and "pushed the limits" as they say. He had sold the businesses he owned and seemed to be settling down to enjoy retirement.

One morning in July of 2017 he went to the local hardware store in his very nice '32 Ford roadster equipped with the very popular 350 CI small block Chevy engine. It was a nice day so he had the top down. When he came out of the store there were four guys gathered around the car, admiring it. Some of the fun and pleasure associated with street rodding is showing off our cars to anyone who is interested, so he gladly answered their questions and pointed out the outstanding features of the car.

When the conversation was over, he got in his car, started the engine, gave it a few high revs and quickly drove off. Apparently showing off to the four guys he just left, he did a "burn out" as he exited the parking lot onto the major four lane highway. He careened out of control across the four lanes, hit a ditch and flipped over. The entire front of the car was destroyed and tragically he lost his life. It was not determined whether he had his seat belt on or not.

The car was inspected later by the Illinois State Police and some local street rodders. It was determined that the accelerator cable was at fault. A very common method of activating the carburetor or fuel injection throttle plates is by way of a multi-strand cable which is inside a housing with fittings on each end that are secured so that the cable will slide fore and aft and thus motion can be relayed from the gas pedal to the mechanism which pivots the throttle plates. This item is made in similar configurations by several manufacturers of street rod parts. Street rodders are usually concerned about the firewall being clean and neat and this system is simple as it is built with just a small fitting on the firewall which accepts the cable housing. Previous to this design, rodders passed a rod through a hole in the firewall. One end of the rod connected to the gas pedal and the other end to the carburetor linkage or to a "bell crank" that had another rod which went to the linkage. This design necessitated some way to cleanly pass the rod through a hole in the firewall and also a seal to keep air from passing through the hole. By the physics of motion this rod rose and fell a fraction of an inch during operation.

The whole idea of the action of either mechanism is to change linear movement to rotary which has always presented a problem. The cable design has helped solve the dilemma except that at both ends of the cable there is still rotary and linear movements such that the cable may go from the top of the fitting to the bottom of the fitting and continually or occasionally rub on it. This wears out both the fitting and the cable in a kind of "hacksaw" movement. Apparently in the case of the crash of my friend, extreme wear was put on the cable which frayed it and then the cable stuck in the housing.

I believe many car owners who use the cable system do not pay attention to setting it up when it is first installed. Care must be made to adjust the bracketry at both the throttle end and the gas pedal end of the housing so that with the throttle at both closed position and fully open position, the cable is not "digging" into the fitting and causing wear.

I have taken it upon myself to casually look at rods at car shows and if I see some undue wear at the fittings, I point it out to the owner and advise him to replace the whole mechanism. They are usually quite receptive to my advice. Perhaps a simple adjustment could have saved a life. The National Street Rod Association is being made aware of the situation and the Safety Teams are paying attention to this important detail of construction.

In the above picture the arrow points to the fitting or adjuster that keeps the forward end of the cable housing steady. In this example the cable has worn through the threaded part of the assembly.

Bearing Down on a Problem
By Dan Tourigny
Alton Bay, New Hampshire

My wife Brandy and I attended the NSRA Northeast Street Rod Nationals in Burlington, Vermont in September of 2016. We pulled a camper behind our '37 Plymouth sedan street rod with the plan to stay overnight at the fairgrounds where the show was held. Our home is on the southern tip of Lake Winnipesaukee, a very big and beautiful lake in New Hampshire.

The trip one way was 150 miles and took 3 ½ hours, some of it on not very good roads. About half way to the show I sensed a mild thumping from the right rear of the car which was rather disturbing. The noise was more pronounced while turning left. The rear end of the car was the ever-popular Ford "nine inch." I wondered if it was brakes, a tire or what. The noise when cornering led me to believe it was a bad wheel bearing. It wasn't extremely bothersome so I "babied" it all the way to our destination.

We enjoyed the show but I worried a little about the car and decided that the trailer should be left at the show fairgrounds just in case there were complications with the car on the way back. The weight of the trailer on the back of the car would aggravate the situation. I didn't need any more problems than what I had.

I got home all right and parked the '34. Later I took another car that had a trailer hitch on it and went back to northern Vermont and towed the trailer home. I later replaced the right rear outer wheel bearing and was good to go!

A total of 14 hours and 600 miles on the two trips up and back is good enough for me to call it a "horror story!"

The Great Flood of 1992
By Roger Dussault
St.Albans, Vermont

The National Street Rod Association has held numerous shows across the nation for many years. The biggest show, the "Nationals," has been held various places but for over forty-eight years it has been held in Louisville, Kentucky at the huge Kentucky Exposition Center. The parking lots and grassy areas around the big exposition buildings can accommodate over 10,000 cars for Thursday through Sunday of the show that is held the first weekend in August each year.

My two friends, Alan and Gerald, and I attended the show in August of 1992. I drove my '37 Chevy coupe, Alan drove his 1932 Ford "Vicky" and Gerald was in his 1939 Ford coupe.

There is a huge complex of buildings where the vendors set up to advertise and sell their products. There was also a food area for many food vendors and other areas for various events going on.

The three of us were inside the vendors' complex on Saturday afternoon when we could hear a roaring noise that became louder and louder. We realized it was raining pretty hard outside. The building was designed with a roof drain pipe system which provided a way for rainwater to pass inside and then outside the building. We could hear the water draining swiftly through these pipes. There were seams in the pipes here and there and water was spraying heavily onto the floor and onto various exhibits. We thought perhaps we ought to get out of the building in case there was some sort of catastrophic flooding about to happen.

As we headed outside, lots of people were headed inside to get out of the rain. We decided we'd better check on our cars. It hadn't been raining for long but there was so much rain in that short time that there was extreme flooding in many areas of the parking lots. Islands were created because of higher ground and water was swirling here and there, headed for drains. Unfortunately, mulch and bark from landscaping headed for the drains as well and it quickly clogged them up, rendering them inefficient. Coolers, lawn chairs and many other objects that would float headed for the drains as well.

Water was 18" deep in places. Some cars were still able to be driven as it hadn't gotten into the engines, so drivers tried to get their cars to higher ground. Some tail pipes were just at water level so created a surreal sight with the exhaust bubbles that were created. One guy on roller skates kept right on skating through about a foot of water. Our own cars were in a pretty good place so that freed us up to help push some cars to higher ground. Many people were panicking and yelling, and directing others about what to do. Many people kept their heads and teamed up to help perfect strangers with their nice expensive street rods. There was talk about what to do if the water got so high that it would get into engines and transmissions and possibly into the rear differentials through the breather holes. People would open doors and water would come gushing out. Wiper blades were used as squeegees to clear the carpets of water.

Trucks came from town with motor oil and provisions for changing engine oil. It was quite a decision for dozens of street rodders to determine just what to do in their own personal situation. Some took a chance that their engines were OK and others didn't want to risk it, so changed their oil.

After raining on and off for about two hours, the deluge stopped and things starting calming down. People had made the toughest decisions about what they should do and adjusted to the remaining water, where the high ground was and whether to change their fluids or not.

All in all, it was a study in human nature as to how people reacted in this unfortunate and strange situation. Many people were very gracious and helpful to others.

It wasn't a life or death situation but it tried lots of folks' tempers and cost greatly in ruined upholstery, hydro-locked engines and other damage. Knowing the nature of street rodders, they probably got back home one way or another and fixed up their street rods with whatever it took. Probably some of them went back to Louisville another year.

Gas Pains
By Bob Brunette
Fletcher, Vermont

During the month of September each year, the Northeast Street Rod Nationals event is held at the Champlain Valley Expo grounds in Essex Junction, Vermont, about twenty to twenty-five miles from our home. Neat treat; cruise to and from the show each day and sleep each night in our own bed.

This story begins during the debugging phase of my newly completed '46 Ford pickup. Of course, I chose to drive it to and from the show even though I knew of a few "bugs" that needed attention. The inaccurate reading of the gas gauge was one of those items on my radar. I was calculating my mileage range based on a best guess of gallons in the tank from fill up to fill up and miles per gallon I could travel on that quantity of gas. This particular day that system failed.

Gloria knew of my gas gauge dilemma and as we passed each gas station, she would remind me of the station we were passing and ask "Aren't you going to stop and fill up?" The male attitude kicked in and I proclaimed I thought we had enough to get us to our destination. We crested the hill just before Bob and Janet Landry's home in Essex Junction, and that familiar cough, cough and sputter occurred. Luck would have it that I had enough momentum that I could coast into the Landry's yard. Immediately I was exposed to spousal verbal abuse, as to why I hadn't listened to her numerous suggestions to stop for gas. The Landry's, who are street rodders as well, had already left for the Expo grounds and I was in deep marital stress.

"What are we going to do now?" was her question. "Well, it's not too long a walk from here to the Expo grounds," I replied. Oh boy, that went over like a lead balloon, but I started walking with Gloria "talking" to me every step of the way. I knew all the while my friends from our street rod club repair shop at the grounds would help me out. Immediately upon arriving at the repair shop, Gloria elucidated all the wrongs I had committed that morning, to all that would listen. Of course, my friends helped. Bob Landry drove me back to his house, got his lawn mower gas can with a long pouring spout, and I put the contents into the Ford's tank. All ended well I thought. I got to the gas station and filled my truck and Bob's gas can. Went back to the Expo grounds and enjoyed the rest of the day albeit with some gentle ribbing here and there.

Not over yet! Every March our club has a banquet at a local restaurant. During the event good natured "awards" are given out. Some are semi-permanent and passed each year from deserving member to deserving member. Other awards are also given out from member to member in the form of hand-made "paper plate awards," which is somewhat of a simple "roasting" of the recipient. When the next similar event came around, the big event of the evening turned out to be a major roasting of yours truly. I was invited to join presenters on the floor of the dining room for the big roast. I think I was the only one that didn't know what was about to happen. In front of my peers I was bestowed with awards in the form of siphoning hoses, a cartoon, a specially painted and pin striped gas can, a fuel pump, a ghost like sheet to cover myself as I used the siphon hoses and a variety of other "gifts" emphasizing my error of the past car season. I have to add that Gloria received a sash to drape over her shoulder highlighting my misdemeanor and a plaque displaying a pair of tiny sneakers, probably as a remembrance of her long walk to the Expo grounds with her ever-loving husband.

Shimmy, Shimmy, Shake Shake Shake
By George Tebbetts
Nashua, New Hampshire

One July several years ago I was driving my '34 Ford three window coupe street rod in a caravan with several other rodders. We were returning to New Hampshire from attending the big Atlantic Nationals car show in Moncton, New Brunswick, Canada.

This was a long trip of 700 miles from Moncton to my home town. Things went well for over half of the trip. Then the car started shimmying and shaking violently left and right but only at certain speeds. It quit shaking about 40 miles an hour so I tried to keep it in that range which was unreasonable at times on Interstates or when the group of cars I was with simply went faster. Of course, I was very concerned and wondered what it might be and what to do about it. I thought of a drive shaft or universal joint problem or wheel weights falling off.

I limped home and put the car away. In a couple of days, I took it to a friend of mine and let him test drive it to get his opinion. We put the car up on a lift and inspected it and found a large bulge on the inside of the left rear tire. Apparently, the belt of the tire had fractured and failed somehow. I then immediately thought about how old all of my tires were. Yup, pretty old. I had been thinking the tread was still good on all my tires but as so many people do, I hadn't thought much about the age of the tires and how the belts and sidewalls may be deteriorating. We rodders tend to put our cars away during the winters and even when we drive them in good weather, we sometimes don't drive a great deal. The tires get old but the tread still looks good. I will keep better track of my tires from now on and put new ones on more often.

Gimme a Brake, Not a Break
By Glenn Turner
Johnson City, Tennessee

I purchased my 1938 Plymouth business coupe street rod in 2001 from a fellow in New York State. The car was pretty nice but needed plenty of work to make it totally acceptable to me. I had a very small garage/workshop at home in Vermont so couldn't do extensive work on it for quite a while. I retired from the big Essex Junction Semiconductor Facility in 2010 and my wife and I then moved to Tennessee to spend our retirement years. One major requirement for our new home was to have a nice big garage and workshop where I would have sufficient room for rebuilding the street rod.

I spent several years putting in a 302 CI, 340 HP Ford engine and a beefed-up C4 transmission with Kevlar clutches and a mild shift kit. I also installed a Lokar shifter, Walker radiator, wheels, tires, an IFS and many small items like power windows, wiring, gauges, etc. I replaced plenty of sheet metal as well. A major project was having a complete upholstery job done by a shop located five miles away.

On March 8, 2018 I took the car for a wheel alignment. The morning of the appointment I awoke to snow in the air and on the ground. Worse than the snow was the fact that there was salt on the roads. I opted to have the car towed to the shop. The appointment was for 2 PM. The tow truck arrived about 2. The driver winched the car on to his flat-bed and away we went. I rode with my street rod buddy, Pete since he wanted to go to the shop as well to watch them align the car. He would then follow me back home in case I had mechanical problems.

It was a short nine-mile drive to the alignment shop. Upon arrival he went to the controls of the bed and I hopped into the car to steer and brake if needed. I pressed my foot down on the brake pedal and it went right to the floor! I was certain all my brake work had been done correctly so this was a very unpleasant surprise. Just then, the driver called out that the car's radiator was leaking. I jumped out, took a look and said to him "That's not coolant, that's brake fluid. You've broken a brake line." He had placed the hook of the winch chain around the front cross member. It had settled right onto the brake metering valve, breaking away both front brake lines.

Of course, there was heavy discussion between the truck driver and myself regarding the snafu. He admitted he was at fault but he offered that the brake lines couldn't be seen as he applied the chain. He offered a reduction to the tow charge and to take the car to his shop and do the repair. Fortunately, it was a good upbeat discussion including an apology from him and even some joking around. I decided to do the repair work myself at home.

We consulted with the alignment guy and decided we'd better fix the brake lines before doing the alignment job. It would have been difficult to drive the car into the shop or even push it safely due to lack of brakes.

At home we were able to roll the car off the flatbed as there was a slight downhill grade toward my shop which allowed the car to gently roll back into my shop. I used the emergency brake as well.

It cost about eight hours of my labor plus $60 in parts to repair the damage. The bleeding of the brakes was pretty difficult to do as well.

As of this writing, the alignment shop is booked up for quite a while, so the time between the first appointment to the alignment shop and the return visit will be about six weeks. Oh well, I have been building the car for 17 years now, what's another six weeks?

Not a good day for sure, but these things happen and we street rodders must roll with the punches. The car will surely be finished in due time.

An Ounce of Prevention
By Don Amundson
Auburn, Washington

I have been a "car guy" all of my life. I started working on cars when I was twelve years old by taking apart a 1937 Nash Lafayette four-door sedan. My grandfather, while not necessarily a car guy himself, was a master mechanic for Great Northern Railroad. He maintained his own vehicles as well as those owned by friends and relatives. As I took the Nash apart, he told me what each part was for and how it worked. At age thirteen, under his watchful eye, I learned how to maintain and service the family vehicles.

Like most teenage boys there were two things we wished for. One of them was to have our own set of wheels. I was fifteen and a half when I bought my first car, a 1942 Chev Fleetline two door sedan. It didn't take me long to nose and deck it, drop the rear end, split the manifold 4x2 and add a ¾ grind camshaft and dual carbs. My first street rod! That was in August of 1956.

I followed the '42 with a '53 Studebaker Starlight coupe, then two more "Studies" and a '56 MGA. In 1967 I bought a '59 Ford stepside pickup and oh yes...got married! In 1977 I bought a '69 Plymouth Roadrunner convertible 383 Magnum which I kept until 1982, when I got serious about street rods. I sold the Plymouth for $4500 dollars, and started looking for a '33 or 34 Plymouth 2 door sedan. In 1982 I located a '34 Plymouth PE two door sedan. In July of that year I bought an '81 Suburban and a car trailer. We went on a five-week vacation which included picking up the car in Louisa, Virginia.

In 1980 I started hanging around with several street rod guys. At that time, I had the aforementioned '69 Plymouth. One of the guys in the group, Fred Kunze, was the Washington State National Street Rod Association Safety Inspector. He talked me into becoming an assistant inspector although NSRA was for pre '49 cars at the time and I didn't have one. As mentioned above, I purchased my 1934 Plymouth in 1982. I put in a Mopar 318 engine backed up by a Chrysler 904 transmission. Now I felt more at home with my street rod friends and also as an NSRA inspector since I owned a car in the category that they sponsored.

In 1981 I replaced Fred as Washington State Inspector. In that year, four of us street rodders from Washington State traveled south from the greater Seattle area to Merced, California to attend the National Street Rod Association Northwest Nationals. We were well into inspection at the site when one of the guys came up and said, "You really need to come and look at this '48 Chevy." We wandered over to the car and he pointed to the ground. We got down and looked under the car. The owner had used a Vega steering box to do his steering along with a bell crank and two tie rods. Lloyd Hall, a member of the Washington State Safety team who was with our group, reached up to the bottom of the box and lightly tapped on the Pitman arm. The arm then lifted up slightly and dropped away from the sector shaft! There was only about a half of a thread left on the lock nut on the shaft when he tapped it so the only thing really holding the Pitman arm on was a slight press fit! The owner, who had gotten on the ground to see what we were looking at gasped and said "good grief." He had driven up from the San Diego area with his wife and two young sons. It truly was the grace of God that he made it to the event without having a fatal accident.

In 1983, at the NSRA Northwest Nationals in Salem, Oregon another "near miss" was avoided. At this time stainless steel front brake flex lines were becoming popular especially for tee buckets, roadsters and other open fender cars. All of the so called "pro built" cars had the "new look" flex brake lines installed on them. It was a Friday afternoon and I was walking from the safety lanes back to where my car was parked in the camping area. Coming up slowly behind me was a guy and two little kids in a street rod. The car was a "period" style Model T with double window windshield, motorcycle front wheels and huge rear tires. He called out to me so I stopped to see what he wanted.

He said that he knew it was too late to get his car inspected in the safety lanes but he would really appreciate it if I could do an inspection for him. He had driven north from Medford, Oregon, a three-hour trip, to the event which was the "maiden voyage" for his new car. I told him to follow me to my campsite. I had him fill out the inspection form and while he was doing that, I started to look the car over. It was a "top dollar" build; a nice example of the style.

We moved down the list of inspection items and came to item #9 which is Steering. At this point we do a "lock to lock" test. I turned the steering wheel all the way to the right and noticed that the flex brake line on the left side was stretched all the way to its limit, so tight in fact that if you plucked it, it would sound like the G string of a guitar! The right tire was tight up against the steel brake line as well. Upon further inspection it appeared that the steering stop bolts that were on the spindles had been removed, perhaps with the idea of grinding smooth the area where they should have been before chroming to present a clean tidy look. It was the same on the right side after turning the wheel all the way to the left. If there had been steering stop bolts, they would have bottomed out on the I beam axle if the flex hoses had been long enough. I had the owner get out and check the flex brake lines while I turned the steering wheel. When I looked at the guy, he was pale. He was almost sick at the prospect of what could have happened should he have been involved in a dangerous turning situation. The flex lines were about two inches too short! He told me he wouldn't leave the grounds with the car until he changed the flex lines. The longer lines would allow the tires to bottom on the frame which was not the best solution but better than having flex lines pull out of their swedges in a panic stop. The next day he looked me up to show me he had fixed the problem for the short term and would make up some firm and satisfactory stops in the near future.

Over the years I have seen a lot of front flex brake line issues. Ninety percent of the time it's been because the flex lines are too short. How does this happen? Think about how many pictures you have seen showing cars at an event, in showrooms or in garages and shops. Also, think how many times you've seen cars in pictures in winners' circles, magazine features and even parked at a cruise-in or other car event. Cars with the front-end sheet metal off during the reconstruction of the car. Engine, front suspension with new calipers, rack and pinion and brake lines all exposed. I will guarantee you that in all of those pictures you will see that the wheels are always parallel, fore and aft. And that, in most cases is the stance of the car when the length of the flex line is calculated, not with the car at the lock point of the steering wheel as it should be!

These two stories are alarming in that we can use our imagination as to what could have happened had someone not come across the two errors in safe building of a street rod. I am glad I was able to be a part of "finding an accident before it happened" as they say!

My '34 Plymouth Deluxe

A Trip to the Market
By Art Stultz
Colchester, Vermont

I got a chance to go to the big annual Hershey, Pennsylvania, Car Show in 1969. We had two cars, the family car and my '32 Ford street rod. I was a school teacher with not much money, so had to build the '32 economically. One thing I did, among many, was to buy a simple set of toggle switches that I mounted in line on a wood panel below the stock dash. I didn't want to drill switch holes in the nice original dash. There were ten identical switches, so I had to memorize which switch was for ignition, lights, etc. I left in the family car early on a Saturday morning and arrived back home about 10 PM. Over a late supper I told my wife how much I had enjoyed the show. Eventually she sheepishly told me that the '32 Ford had been in a small "fender bender" while I was gone.

Some groceries were needed and I had the family car. A man had driven a bit too far ahead while stopping for a traffic light, and while backing up had run into the '32. I went to the garage to assess the damage. The left front fender was crumpled a bit. My wife said she had the phone number of the other driver, so assuming that everyone involved would agree that he was at fault I called him the next day. He said he was very sorry but added "The young fellow driving the car said he didn't know how to turn on the lights, and since it was getting quite dark, I didn't see his car behind me while I backed up." I thought to myself: "The young fellow..." Must be the young fellow was my stepson who hadn't had his driver's license for long. I told the man that I would not pursue his paying for damages. I thought this whole thing out and realized that my wife had cleverly never said she drove the '32 and didn't say who did! I confronted both my wife and stepson, and he explained that since there were ten identical toggle switches on the under-dash panel he didn't know which switch was the one for the lights. My retort was that this was just one of the many reasons I had warned him not to drive the street rod without permission.

I straightened out the crumple with lots of hammering and the Bondo routine and got it back to original shape. Lesson learned: Install a secret hidden master shutoff switch!

A Bent Cam Pin
By Everett and Jane Wrightington
Rochester, Massachusetts

We drove in our fiberglass '34 Ford coupe to Lake George, New York in the early fall of 1989 to attend a big car show. This was the first year on the road with this car.

We had had a very good season with it and had put about 3,500 miles on it at that point. The engine was a 350 CI Ford Cleveland motor. As I was putting this engine together, I wanted to "advance the camshaft" just a few degrees, so I drilled out the cam pin hole in the cam gear by hand and put an offset ring back in the hole to give the cam about three degrees of advance. That little Ford ran great for all of the first 3,500 miles. We had just had breakfast in one of the small off the beaten path restaurants in Lake George and were driving through the town when suddenly the engine seemed to "load up" on me. I thought maybe a float had stuck in the carb and it was getting too much gas. We found a spot in town to park it and let it cool down a bit before I worked on it and looked around for our pals that had been travelling with us. When they came to help, I told all of them about our problem with the car and that it was no big deal and we would get to it after we shopped and had some fun in the "touristy" village of Lake George.

Well, it came time to check out the car. We lifted the hood and then I got into the car while one of our pals looked at the motor as I turned it over. He told me to stop cranking, and then pulled the distributor cap off. He then told me to crank it some more. Then I heard the bad news that the rotor on the distributor stopped for a few seconds (Was this possible?) but was now turning again. We put the cap back on and tried to start it up. Well, it did start, and I said that we were going to try to make it back to the motel. We made it back all right...a short trip of about three miles.

I knew that this car would never make it back to our Massachusetts home in this condition so we had a good friend come out with a trailer on Sunday to pick it up.

Once we got the Ford home, I started to take the engine apart and found the pin on the cam had broken off which let the gear spin. I had bent both valves on the number one cylinder. By the time we paid the expenses for our friend coming all the way from Massachusetts to get us, then getting us back home, his time, food, gas and then the cost for the repairs, etc., it turned out to be a very costly trip.

We were left with two things to think about:

#1 How did the ailing motor run well enough to get us back to the motel?

#2 I had drilled the hole in the cam gear by hand which surely was not a perfect job, by any means. So...how come it ran so well for 3,500 miles without trouble?

We are still in that little Ford with about 58,000 miles on it. We hope to see you on the road sometime. Be safe.

P.S. We think back to all the "rod runs" that we have attended and been a part of; all the good times we had and all the great cars we have seen. There is one thing that stands out above and beyond all those things: it's all the good people that we have met and become friends with over the years. God bless.

Leaves in the Dumpster
By Dennis O'Brien
Charleton, Massachusetts

My wife and I have been to lots of car shows over the years to promote my business, "O'Brien Truckers." Sue and I shared driving our '34 Ford panel truck street rod which is loaded with the display items that I sell.

In 1987 we were passing through Kansas City, Missouri on Interstate 35 on our way to Carlisle, Pennsylvania to attend one of the Goodguys shows. My wife was driving. We both could sense something was wrong with the suspension as the truck started swerving left to right almost uncontrollably. We were close to Exit 17 which had some construction going on, so she pulled behind some temporary orange safety barrels and found a small area where we could park.

It was popular in those days of street rodding to take the front springs from an early Chevrolet truck and use them on the rear of a street rod. These front springs were longitudinally placed, one set on each side so they were pretty stable. Our truck had a set of '53 Chevy front springs on the back. A leaf of the spring on the right rear had broken.

We got out our road atlas and Fellow Pages which we always carry with us and started walking down the ramp of the exit to find some help. Right away, a pickup truck with lettering advertising dumpster trash service stopped next to us. We told the driver what had happened with our truck and he said he owned the trash company and could help us out. We rode with him to his place of business where there were big shops and lots of trucks and dumpsters on the property. He got us a tool commonly called a "come-along" which is a kind of a block and tackle unit with two hooks that can be used to lift loads or pull things together. We thought it could be used to hold the spring set together enough to drive the truck safely to a spring shop that he knew about.

Back at our truck I raised the right rear with a jack I had with me and placed one hook of the come-along around the bracket that supported the front of the springs. I hooked the other end of the come-along around the shackle at the back. This way I could draw the rear toward the front to keep the whole set of leaves from losing alignment and support. The next leaf up from the main one had a "curl" or "half-moon" on it which was part of the way the set of leaves was kept in alignment and I wanted to make sure I kept it in its proper position. I was able to drive the truck slowly to the spring shop that the fellow told us about. We then went to a motel in West St. Louis where we spent the night since there wasn't much time left in the day to start on the repair. In the morning, we drove judiciously to the spring shop, which fortunately had just the spring I needed for the repair. I did the replacement right there in the yard of the spring shop as I had just enough tools and jacks with me to do the swap and then we went along our way. The trash company owner said to keep the come-along!

We felt pretty lucky that there was someone willing to take time from his busy day to help us out and we were glad that we had brought enough of the right tools to do the job ourselves. It doesn't always work out that you have the right tools with you, as it is impossible to know just what, if anything, will go wrong with your street rod.

Old Time Street Rodder
By Art Stultz, Colchester, Vermont
Roger Kendrew, Milton, Vermont

The late Ken Bucklin was an "old time street rodder." That is, he liked to build his cars for speed and power and drove with reckless abandon. He had little concern for safety that we are concerned with in this day and age of safe cars and safe street rodding.

He had lots of personal and business connections over the years with street rods, trucks, RV's and race cars. His favorite street rod probably was his '34 Pontiac two-door sedan equipped for speed and power with a big block Pontiac engine, about a '61. He ran fenderless and without much regard for speed limits, safety or defensive driving. His car was never finished as he was not concerned with upholstery, nice seats or a pristine paint job. The side windows were made of scrap plexiglass. As long as it would go fast, he was happy.

One day I ran into someone who knew I was a "car guy." He asked if I knew who drove "an old, white hot rod car with blue flames and no fenders." He said, "I saw him going like hell on Spear Street the other day." Oh yes. I sure did know him. That would be Kenny Bucklin.

Ken had a long history with everything to do with cars. He had worked off and on with many race car "pit crews" as he was an excellent mechanic, especially with improvising as he went along. Making brackets, supports, sheet metal patches and things with odd pieces of metal he had hanging around was his forte. He could knock metal with the best of them as long as you didn't want it detailed and completely finished.

One day he needed a shaft to go between two steering u-joints on his '34 so found "just the thing": a long 3/8" drive extension from his tool box! Welded it right in between the two u-joints! Good to go!

Five cars from our street rod club planned to take off from his house one day at six in the morning headed for a big car show in Worcester, Massachusetts. He had his '34 supposedly all ready to go, but when he turned the key to crank it the dreaded "click click" came from the starter. With little concern he simply whacked the starter with a hammer and off it went on the next turn of the key. I thought we would have to wait until the auto parts stores opened to get a new starter and put it on but no...he simply said "OK...Let's go!" and off we went. That was Ken.

In 1984, a half-dozen of us who belonged to a car club headed off from northern Vermont to Columbus, Ohio, to attend the big NSRA Nationals. Ken hadn't taken the time to change an ailing alternator on his Pontiac so simply brought along a battery charger. Each evening at the motel he took the battery out of the car, brought it inside his room and charged it all night. Next day he put it back in and drove for the day. This obviously worked as the engine always cranked well. On the same trip one of his tires had a big ugly knot on it which gave him a somewhat bumpy ride but he didn't mind. Do you suppose he had a spare tire and a jack? Why no!

Each January there is a giant "swap meet" in Springfield, Massachusetts. Ken had a big RV at the time and one year volunteered to take seven of us street rodders in this rig to the meet and back. We soon learned that he didn't take care of his RV any better than his street rod. We hopped into his RV and headed down the Interstate. Since it was pretty cold out on this winter day, we anticipated that soon we would be warm and toasty when the heat came on. It never did! He said, "Oh yeah, I forgot to tell you guys that I don't have heat in this thing but I can turn on the stove burners." That didn't help much so we just toughed it out. On the way back he had some doubts whether the alternator was putting out enough, so we removed the "doghouse" engine cover and managed to do some testing on the fly up the Interstate. Lights were a bit dim but we carried on. About 40 miles from home the engine started bucking and jumping, and he announced that he had been trying to find a gas station for the last dozen miles but couldn't find any. We had to pull over and stop. A couple of guys going to Canada stopped to help us out. They happened to have a full gas can with them. What luck! It got us to the next gas station.

The car club Ken belonged to put on a weekend "Rod Run" for many years at Lone

Pine Camp Sites in Colchester, Vermont. He always helped out with preparing and serving food amongst other jobs. He had to go to a convenience store which was two miles up the road to get something he had run out of, so asked Brian, a fellow rodder in the club, if he would like to go with hm. Brian said yes, he would. Ken asked if he had ever driven this car and Brian said he had not. "Would you like to?" he asked. Brian said "Umm...sure." Brian slid into the driver's seat and off they went. Brian was very respectful of Ken's car so drove carefully for the first half mile or so. Ken said, "Hey! Get on it! You don't have to baby this thing." So, Brian "stepped on it" and thus witnessed first-hand the torque and power the big Pontiac engine had. Ken was surely enjoying it all but then had a sudden remembrance and announced: "Oh...Brian...I forgot to mention that I have brakes only on the left front of the car so take it easy up here at the corner." This scene fortunately ended without incident as Brian stopped the car okay but was a bit shook up.

Later on in life Ken had a connection with Florida. He and his extended family liked to go there for vacations now and then and eventually settled permanently in the Daytona area. On one trip back to Vermont from the Sunshine State his old family car started to give him some trouble. The crossmember that supported the rear of the transmission had somehow rusted to the point of collapse, which made for heavy vibrations as the driveshaft u-joints were at severe angles. He pulled back the floor carpeting and cut a square hole at the top of the transmission tunnel with a big hammer and chisel. He found a two-by-four and placed it so it straddled the hole, ran a chain around the transmission and up over the two-by-four. He raised the transmission up with a bottle jack, tightened the chain and ran a bolt through two of the links. This very "shade tree" repair got him back to Vermont. Don't ask me where he got the hammer, chisel and chain. He probably had them in the trunk.

Ken lived a charmed life as far as we could see. As far as we knew he never got into an accident or got stopped by the cops for speeding. But one day in the late 1980's it all caught up with him at the Oxford Race Track in Maine. He drove his '34 most everywhere so drove it there as well from his home in Vermont to a big race that weekend. He worked in the pits with the Dale Shaw Racing Team throughout the evening. When the racing was over, he was leaving the parking lot probably with the "racing mentality" still in his head, and getting on the gas pedal too hard as usual ran headlong into a big post in the parking lot! He tore up the front end of the '34 pretty badly with lots of damage to the radiator, front frame horns, fenders and tie rod. The race car team was still there on the grounds so Ken, along with his brother-in-law and race crew guys, worked for a long while with the tools and equipment they had and got it going again. Ken of course drove it back to Vermont that night!

Ken and his family moved from Vermont to Florida in the mid-1990's where he got a job in the Florida State Penitentiary supervising prisoners who maintained state cars and trucks. Leukemia finally caught up with him and he passed away.

My Shoes are Showing!
By Dennis O'Brien
Charleton, Massachusetts

My business, O'Brien Truckers, sells aluminum plaques, belt buckles, air cleaners and many other street rod accessories at car shows and by way of the Internet. It is important that I go to as many shows as I can to display and sell my wares.

About 1995, my wife and I were driving during the late evening through Parsippany, New Jersey, in my "company truck," a 1934 Ford panel truck street rod. Our daughter and her husband were behind us in their Cadillac Escalade that was loaded with lots of my items for sale. We were on our way to York, Pennsylvania, to attend the big NSRA Nat's East weekend car show.

At 11 PM we exited the Interstate highway and decelerated to a traffic light that was red. When the light turned green, I stepped on the gas pedal and...nothing! It was like I was in neutral! I got out and looked under the truck to determine if a u-joint had come apart or just what had happened. I noticed that the right rear wheel had shifted laterally. There was a gap of a quarter inch or so between the backing plate and the brake drum that allowed me to see the brake shoes! Eeoww! It looked like the axle had broken and was on its way out of the housing.

The traffic light was changing red to green and back again and traffic was piling up behind us, so we pushed the truck through the intersection and into a gas station. That is all it was, a gas station, with the usual pumps and a kiosk where an attendant took the money. I asked the attendant about the possibility of getting my truck fixed there and in broken English he said he was just selling gas.

There was a pickup parked at the gas island nearby. The truck had a plaque on it as well as emergency lights that indicated he was with a fire company. Perhaps he was also a "car guy" of some sort or would have knowledge of the area and where we could get the truck fixed.

He spotted us in our street rod and immediately took an interest in what was going on with us, so we discussed our breakdown with him. We told him we thought maybe we had broken an axle. He surely was a car guy as he said he owned a '56 Chevy and also a shop where he did custom car work. Yes, he knew the area and had a phone number of someone with a ramp truck that could pick us up and take us to his shop.

At his shop we determined that it was not a broken axle after all, but the main right lateral axle bearing had failed to the point of losing its tight fit and allowed the axle to shift away from its normal position. I drive fast but not real hard so thought that although it looked like a broken axle at first, I doubted that it was.

The rear end was an 8" Ford Mustang. Right there in his shop was a hot looking late '60s Mustang with a 429 cubic inch engine. The fellow was putting in a new Mosher rear end in place of the original one which he had already pulled out and had stored against the wall. We pulled out the axles from the old rear end. They both fit my housing so we made a deal with the fellow and installed them.

About one in the morning we were done and good to go! This whole affair was a bit of bad luck but with some good luck as well as we came across the fireman just at the time we needed someone and he was willing to take us to his shop, provide the axles that we needed and spend the time to get it fixed. A true "Brother of the Wrench."

Not Dead on the Road Yet
By Fred Hout
Altamont, New York

I am a lifetime member of NSRA so I really enjoy being involved with rodding and going to the annual Northeast Street Rod Nationals in Burlington, Vermont.

In early September of 2015 I was getting my 1940 Packard hearse street rod ready to make the trip to the show. I always enjoy driving it and showing it off as it is pretty unique as a hearse and as a street rod as well. This check-up was done about two weeks before the event, so I had given myself time to take care of any minor problem with the car. Sure enough, the engine, a 460-cubic inch Ford V8 was not running well enough to give me confidence to set out on the three-hour trip.

I thought perhaps the problem resided within the Holley four-barrel carburetor so I worked on it for quite a while and determined that it was probably the vacuum diaphragm for the secondary throttle plates, so I replaced it. I also put in diagnostic time on the electric fuel pump which was very hard to get to but I finally decided it was not at fault.

My wife and I set out for the show thinking all would be okay, but as I got up to speed on the highway, I felt the engine just wasn't running quite right. I kept on going however, as I could live with it, but at about the 80-mile mark away from home the radiator sprang a leak. I couldn't ignore this of course, so had to give it all of my attention.

We had been to the show many times in the past and had an overnight arrangement with friends of ours in Granville, New York, which was about halfway to our destination. We would stay there overnight and continue on to the show the next day. As it worked out the radiator leak happened about five miles from his house, so I limped there even though the leak was severe.

The problem was obviously a crack in the area of the filler neck and overflow tubing. Fortunately, my friend had the tools, equipment and expertise to clean the area well and solder it up.

We continued on to Burlington the next day and enjoyed the show once again. The hearse gave me a hard time but never died!

Hot Shot Studebaker
By Ken Bessette
Williston, Vermont

I owned and operated my own body shop for many years and finally was able to retire. But like so many who give up their careers, I felt I should go back to work at a job that was not too demanding. I thus took a job delivering parts for Bond Auto Parts in the Burlington, Vermont area.

I made repeated deliveries by a house and yard in the town of Williston and would see a car covered by a blue tarp that sagged down a bit so I thought it must be a convertible...maybe an old one. Very interesting!

One day I once again passed by this same yard and this time the tarp had partly blown off so that I could see it was a 1953 Studebaker Commander convertible.

I later learned that one of the several previous owners had made it into a "shorty car" by removing about 34 inches from the frame and body. He had sawed the car in half just behind the front doors, removed a section of body and frame and re-welded them together.

I took a liking to the car more and more and wondered if the car might be for sale. I finally stopped one day and asked the renter of the house who the owner was and thus was able to contact him. He gave me a price but said he really didn't want to sell it as he had plans for continuing the build some day. A story we have heard many times from lots of "car guys."

Three years went by and each time I passed the place the car was still there. I gave the fellow lots of time to change his mind before I stopped once more to ask the same question. This time he said yes, he would like to sell it, but now it was at a higher price than before! I bought it anyway and actually drove it home a few days later. It needed lots of work.

The car had a 1964 Pontiac 389 cubic inch engine squeezed into the engine bay. This big power plant would have plenty of power whenever I got it finished.

I worked on the car and engine for several months but couldn't get it running satisfactorily. A good place to start was the cranking circuit so I called a friend who came over to give me a hand. Between the two of us we tried to square up the starter where it bolted to the engine. The starter was cocked over in an odd position so the starter drive gear wouldn't mesh with the ring gear of the automatic transmission properly. The car had the original steering, which being a Studebaker had some very unique configurations of crossmember, steering box, Pitman arm, control arms, etc. Between the stock steering and suspension plus this big Pontiac engine, the starter, which was located on the left side of the engine very close to the Pitman arm just barely fit in there.

Before getting the starter in straight we had to first square up the transmission onto its crossmember. It took some prying and shimming to get it straight and where it ought to be. Now on to the starter. It had to come out first to fix the solenoid cover and wires that fastened to it that were pretty much cobbled up. Back in with the starter, which was quite a squeeze, but we made it. We then tightened up everything related to the job.

OK…now…time to see if it would crank all right. The battery was in the trunk with long cables running forward for hot and ground. In the engine compartment in series with the hot side cable was a master disconnect switch. Thinking all was going to be just fine I threw the lever of the switch to the "on" position and "pow!" Something made a loud popping noise at the back of the car and smoke poured out of the trunk for a few seconds. I quickly turned the switch back off.

Going back to take a look at the battery, heavy current flow had instantly heated up the cable connection to the battery negative terminal and blew it apart, melting it totally into an open circuit. Examining the whole circuit to see what had happened we discovered that the large threaded post of the starter solenoid where the positive cable was attached was just barely touching the left motor mount. There was plenty of gap before we straightened the transmission and starter but now there was an almost imperceptible contact between post and ground. We were done for the evening as there didn't seem to be any really good quick fix for this dilemma.

This was a good lesson regarding the master disconnect switch. This switch surely has a place in the scheme of things as it is a convenient way to turn off all the electricity in the electrical system. However, when electrical work is to be done the mechanic should disconnect the ground cable of the battery. When all work has been done make the ground cable the last thing you attach. Place a couple of shop rags soaked in water on top of the battery in case of battery gassing. Make the attachment by lightly tapping the cable one time on the post to see if there is a spark, large or small. The spark means there is a complete circuit to ground somewhere in the system that needs attention. A heavy spark means there is a serious and dangerous ground of the hot side. Better yet disconnecting the ground cable at the engine assures that the spark, if created, will be farther from the battery. Batteries are known to blow up when sparks are nearby and the battery has been gassing.

As time went on and many other problems with this rig reared their ugly heads I decided to get rid of the big Pontiac engine in favor of a "good old small block Chevy." I trashed the ugly original Studebaker steering and suspension in favor of a Mustang II suspension which proved to be much better.

All is well now after lots and lots of work. The shorty "Hot Shot Studebaker" drives and handles pretty well!

My "Shorty" '53 Studebaker convertible

Saving Buddy
By Art Stultz
Colchester, Vermont

I taught high school Automotive Technology for 27 years, six in Maryland and 21 in Vermont. There were many satisfying and pleasurable aspects of my job. The main one was that I was teaching something I had a very deep interest in. I had always been curious about mechanical contrivances and thus as a youth had built and repaired model trains and airplanes, bicycles, go carts and finally automobiles. This interest led me to obtaining a college A.A. degree in Industrial Education with a major in Automotive Technology, followed by securing a teaching job at an area Vocational Technical school in Maryland from 1964-1970.

I enjoyed lecturing in the classroom and presenting demonstrations as well as conducting the shop "hands-on" work very much, and appreciated meeting and becoming friends with so many young men (and a few young women) who were interested in the vocation of Automotive Technology.

I had acquired a '32 Ford street rod previous to the start of teaching, so of course showed it to my students and had them work on it once in a while as part of the hands-on aspect of teaching.

In my fourth year of teaching, I came across and bought a '55 Chevy which was in pretty rough shape. Beat up body, bald tires, no engine or transmission. This was an excellent opportunity to "kill two birds with one stone" - and teach some good basic hands - on with the car and get it running at a pretty low cost to me as well. The plan was for the students and myself to get it fully operational, and it would be my stepson's car as he would have his driver's license by the time it was finished.

I had very little money in those days so looked for any way to save money. A student gave me a 265 CI engine and another student came up with a Powerglide transmission that was wasting away in his yard and he didn't mind donating it to the cause.

The class rebuilt the engine, mated it up to the transmission and put it into the engine bay. After the lesson on the basics of properly putting in the distributor and wiring the ignition system, it was ready for the big day.

We then attempted to start it. An engine coming to life for the first time was always a major event for all the class members, so I allowed them to drop their work assignments in other parts of the shop and gather around for the potential start.

I taught plenty of safety in all aspects of work in an auto shop atmosphere and was doing the same in this situation. We had previously set up the brakes. A student was behind the wheel to turn the key and operate brakes if need be. The transmission was supposedly in neutral or park.

One of my students was a fine young man named Matthew whom we called "Buddy." Buddy was one of several students gathered around to see the "fire-up." He was standing directly in front of the car maybe a half foot away from the bumper. Ahead of the '55 was a '58 Plymouth that was donated to the school and that we hoped would be a contender at the local drag strip when it was all set up.

The student behind the wheel commenced to crank the engine after someone poured a little fuel into the carburetor. It cranked and cranked but no action. More priming and more cranking and suddenly it came to life.

The engine roared and the car instantly leaped forward! The transmission was obviously in drive and not in neutral or park! The fellow behind the wheel had the sense to instantly slam hard on the brake pedal. The car stopped a split second after it had lurched forward and I could see that Buddy's legs were pinned between the bumper of the '55 and the '58. Buddy exclaimed "I'm okay...not hurt!" I rushed to see just where his legs were and there must have been no more than an inch or so between the front of his legs and the '55 and the back of his legs and the '58. Phew! What a relief! Surely if things had been a little different both of his legs would have been badly broken. My main interest at that point was Buddy's injuries if any, with a thought in the back of my mind that I might be out of a job very soon!

Looking into what had been done...or not done...I hadn't even thought about wiring in the neutral safety switch. We hadn't blocked the wheels with wheel chocks. I was too casual about checking if the transmission was in park.

This was an excellent safety lesson, especially to show the need for an operating neutral safety switch. The National Street Rod Association Safety Program pays plenty of attention to this switch and tests it in a safe manner as well. This item is in the program's top three of having the most defects when the safety results are summed up by NSRA at the end of the street rod season.

Needless to say, I have made sure all of my street rods have their switches working properly and I test them often.

Looking back, I am very happy things worked out the best for Buddy!

The Grasshopper
By John Smith
Cohoes, New York

I owned a 1937 Pontiac street rod from 1980 to 1994 and decided to sell it, so I took it to the NSRA Street Rod Nationals East in York, PA in 1994 in hopes of finding a buyer. There I sold it to a Michigan couple who was also attending the show. Part of the deal was that I was to deliver it to their home in Milford, Michigan.

We street rodders can't go long without a street rod in our possession so I also had arranged to buy a 1938 Chevrolet two-door sedan street rod. This car was owned by a neighbor of mine here in New York State but the car went with him to Ortonville, Michigan when he moved there. I reunited with him at his son's college graduation nearby my home in New York State in May of 1994. We got to talking about the car. I offered to buy it and he agreed to sell it, but I explained that I had to sell my Pontiac first. He gave me six months to sell my car. The show in York the next month provided an answer to the need to sell it quickly.

Among other features of the '38 was a graphic of a grasshopper cleverly painted on its "beltline" which I thought was cool.

Since the sale of the Pontiac and the purchase of the Chevy both involved Michigan, the plan was that my wife and I would drop off the '37, then pick up the '38 and drive it back home.

On the way to Michigan with the Pontiac, I stopped for fuel at a gas station. When we resumed the trip, the engine wouldn't crank as the battery had apparently "died." The battery seemed to have been overcharged. (Perhaps gassing of the electrolyte?)

I diagnosed the problem as a bad regulator. It was one of the early style alternators with the external regulator. I happened to glance down at the running board and there was a dead grasshopper! Was this a bad omen of some kind or what?

I got a jump start by way of a stranger who happened to be at the station and proceeded the rest of the way without shutting off the engine for fear of not being able to start it again. The trip was not as much fun now with the worry about stalling or the prospect of a pressing need to shut the engine off.

I finally pulled in to my destination which turned out to be a nice old Victorian home. By then my car was covered with dirt and dead bugs but no more grasshoppers, thank goodness!

Whew! Made it! My wife and I enjoyed an evening in this nice mansion with our new-found friends relaxing with champagne, cheese and crackers and lots of laughs about the ominous omen of the dead grasshopper.

We stayed overnight and the next day the couple drove us to Ortonville, 27 miles away. We spent that night with our old friends and then drove "The Grasshopper" back home with no more problems. We had gotten the jump on the bad omen!

Lobsters Paid the Bill
By Ed Miller
Spring Hill, Florida

In August of 2004, my wife and I were on our way to the NSRA Nationals in Louisville, Kentucky. I was driving our '34 Chevy street rod in the Wheeling, West Virginia area when something obviously broke on the front of the car.

I was able to pull over to the side of the road and check it out. The front suspension was the typical Mustang II design with the long and short control arms. The bracket that held the lower control arm on the passenger side had completely broken off.

Despite the damage, I was able to drive slowly and carefully to a motel and from there used my trusty NSRA Fellow Pages to find a street rodder who lived in the area. The fellow, who was an FBI agent, gave me directions to a welding business close by. Another street rodder, my good friend Bill Roberge, was traveling with us. He was able to take me to the welding shop to see what they could do for me as well as drive me back to the motel where we picked up the car and drove it to the welding shop.

The next day a mechanic rebuilt the control arm, which took four to five hours. When we were done, we used the simple "string method" to check and adjust the toe-in of the front end.

Being a "Brother of the Wrench" and a street rodder himself he didn't charge me anything!

When I got to Louisville, I checked his work on the control arm and found it to be an excellent job. I checked the arm on the other side and it was OK too.

I wanted to thank the welder for his fine work so the next Christmas I sent him a dozen lobsters. I was quite sure he would enjoy the fine delicacies as I do. The crustaceans were frozen and packed in Styrofoam packets so made it there in good shape a couple of days later. Not being a New Englander like I was at the time, he called me and asked: "Thanks a lot Ed, but how do you eat these crayfish?"

I Shoulda Gone to Church
By Stan Morrow
North Ferrisburgh, Vermont

I was bitten by the "car bug" at a real young age. I was brought up on a very rural farm in Vermont and my Dad had me driving cars, trucks and tractors around the farm and even with him on the public dirt roads around home as a mere lad of six and a half years old. There weren't many other vehicles on the roads in those days in that area and police cars were very rare as well; there wasn't much chance of getting caught driving at such a tender age and of course without a license.

The farm had lots of room with barns, garages, sheds and the like where I could store and work on the vehicles that I acquired.

In 1961 when I was 14, I traded a Whizzer Motorbike for a 1930 Plymouth coupe. Since I wanted a roadster, I cut off the top with an air chisel. Someone had put a Chevrolet six cylinder in the car but I was into Ford flatheads so I came up with a '47 Ford engine that I put in it instead. It had a three-speed manual transmission behind it. I found what I learned later was a very popular nine-inch Ford rear end which probably came from a '57 or '58 Ford. My cousin helped me weld up a driveshaft and the proper u-joints to make the connection from transmission to rear end. I got it running and drove it around the yard a bit.

The next job on it was to paint the frame. I had an electric grinder with which to prepare the frame but there was no electricity in the barn where I did most of my work. One Sunday morning my family wanted me to go to church with them but I passed it up as working on the '30 promised to be much more exciting. I drove the car from the barn to the house which was about 200 yards away and where I would have electric power. I was well into the grinding when Mike, my twelve-year-old friend and neighbor happened by and volunteered to help me. He immediately asked me how I got the car down to the house. "I drove it, of course" I replied. Since he had known about the car and its construction, he did not believe that I had gotten it running. He kept pestering me that he wanted to hear it run but I was busy with my project. With further cajoling I consented and we agreed to go for a short ride down the dirt road and back. I knew full well I shouldn't have been going on the public roads with both myself and the car not being licensed, not to mention Mike being with me but I figured we wouldn't be gone long and would simply be back in no-time, safely into the yard where we started from.

There were no seats in the car so I had rigged up a simple wooden box to sit on as I drove. There were no floorboards either so Mike just stood on the frame and braced his arms on the body. I started the engine, pulled away from the dooryard and "floored it." This surprised Mike greatly as he wasn't hanging on nearly well enough, so he flew backwards and landed just under the rear "package tray" with his head near the drive shaft. I slowed down so he could get up and support himself as we continued our short journey. He wasn't hurt at all.

I was driving south away from home and turning left onto the next road and up a small hill when the engine started overheating. I was not surprised as the radiator was an old one that I had come across somewhere and just put it in without much thought as to condition or coolant capacity. The steam pressure overcame the cap and poured out of the overflow tube and was coming straight into my eyes. The brakes began acting up as well.

Meanwhile my mother, father and sister had come back from church. My mother had spotted my car at a distance just when I was about to make the turn. She left everybody else at home and proceeded to come after me in her car, a '61 Dodge Dart. I had no idea she was back there when the overheating and brake failure commenced and as I took a quick U-turn and headed back home. I came up over the rise in the road I had come over before and there was my mother out of her car and waving at me with both arms trying to get me to stop. With my wimpy brakes and steaming radiator, I wasn't about to stop but just steered off into the ditch and around her and her car and headed home.

Of course, when we both arrived at home, I received a severe tongue lashing and probably got some privileges taken away but after fifty-seven years I can't recall the details.

I brought the car back up to the barn and parked it. I worked on it now and then but with many other projects in my life the 1930 Plymouth never became a full-fledged finished street rod. I still live in the same house as I did then and still have the car, safely stored away in the barn.

A Hauler Horror
By Charlie Bryson
Ursa, Illinois

I have been the National Field Director of the National Street Rod Association since August of 2013. This position entails working at many shows across the country throughout the street rod season. I do find time to serve as part of the crew of a Late Model Dirt Track Racing Team with a very good friend of mine, Rick Frankel.

On Friday night, June 11, 2016 our team had won the feature race at Tri-City Speedway in Granite City, Illinois. Our race car hauler was a 2002 Freightliner with a 24-foot living quarters section and almost all the conveniences of home. We usually stayed overnight in the truck at the race track on Friday nights and then would drive the next day to the Saturday venue. The Saturday night race on this particular weekend was in Pevely, Missouri at Kenny Shrader's race track, Federated Speedway I 55. The Friday night races got over a little earlier than they usually do so we decided to go back home to the Quincy, Illinois area rather than stay overnight at the track.

Saturday morning, we gave the race car a good inspection, refueled the truck and headed south to the I 55 race track. We had gone about 14 miles down the road when we heard a loud explosion. The right front tire blew out on the hauler which was pulling our 38 ft. long enclosed trailer with the race car inside. At about 65 miles per hour the truck took an immediate right turn, careened off the highway and hit a 16 ft. rock wall head on. The hauler and the trailer were totally destroyed. The race car was not damaged much, although on impact it tore off all the straps that held it down and traveled forward all the way to the front of the trailer, almost to the fifth wheel area.

I was sitting on the couch in the living quarters of the hauler. I heard Rick, the owner/driver of the truck yell that he could not steer it. The next thing I remembered was waking up on the floor with Rick's wife lying across me screaming that she was dying. I assured her that she would be okay, not realizing that I had an eight-inch cut on my head. It ran from the front of my head, up and over my ear to the back of my head. Rick took a closer look at me and thought I was dying too!

Rick's son, Rickey, the driver of the race car, had been sitting behind me on the couch and was now lying on the floor with a severely broken collar bone. Rick's wife, who had been sitting in the passenger seat had her foot caught up under the dash.

It was a busy area so many people came upon the scene and someone obviously called 9-11 which brought police, fire equipment, an ambulance and a helicopter to the scene in a short time.

Rick's wife's injuries were severe enough that she was Life Flighted from the scene by helicopter to a hospital in St. Louis. She had her heel completely broken off, a broken ankle and a four-inch section of bone broken out of her leg below her hip. She still has pain and discomfort two years later.

Rickey's broken collar bone required 15 bolts to repair.

A half hour after the accident I was taken to the hospital in Hannibal, Missouri, only ten miles away. My head was sewn up and I spent the night there. The next day they brought me by ambulance to Quincy, Illinois. The hospital there checked me over and sent me home; what a mistake on their part! The next day I was in so much pain in my neck and arm I had to go back to the hospital. They readmitted me and proceeded to do tests. Later that evening they tried to do an MRI but I was in so much pain I couldn't stay still long enough. They took me back to my room and started giving me massive doses of pain killers to stem the agony. The next day they were able to do the MRI the way I wanted them to. Later that afternoon the doctor came in to talk to me and advised me that I had the worst ruptured disc in my neck that he had ever seen. He wanted to operate that day. I'm sorry to say my family had issues with this doctor and told him there was no way he would operate on me and that I wanted another doctor that I knew of. However, the doctor we suggested was out of town and wouldn't be back for three days. My brother expressed his thoughts to the doctor that there was no way he would touch me and told him to leave my room and not come back. We would wait for the other doctor.

Three days later I was operated on. They went through the front of my throat and removed the disc and replaced it by screwing in a titanium butterfly that fused two vertebrae together. After a lengthy stay, I was released from the hospital with the only reminder of the accident being partial loss of the use of the thumb and index finger of my right hand.

We now have a replacement hauler and trailer. Rick paid $59,500 for a 2012 International with a 30-foot living quarters. The old truck had seat belts for just the driver and passenger but this one has belts at every sitting location in the vehicle. The cost to replace the trailer was $39,500. Insurance did not pay for everything so it was a pretty costly mishap for sure. We are back to racing again and winning feature races at many tracks around Illinois and Missouri.

One thing that I would suggest to everyone is that if you have the chance to sign up for Air Lift in your area please do so. Rick's bill for the helicopter was over $32,000. I paid the membership fees for the Air Lift for both Rick and Rickey's family for Christmas presents last year.

This was a good lesson in the need for paying attention to all aspects of the safety of a vehicle whether it be a street rod, car hauler or a trailer. Just one tire can make all the difference in the world.

New is Not Always New
By Jed Greeke
Dover, New Hampshire

In 2012 I purchased the car I had always wanted, a rare 1972 Buick Skylark Sun Coupe. Numbers were never kept during production but the consensus is (as published by Hemmings Motor News) that about 1800 were built. I did restoration work with the help of a local shop. I did it in steps, completing what I could myself and paying the shop for what I could not do. The last work the shop did was body work and paint in the summer of 2016, and then I upgraded the front brakes by changing them from drum to disc. The restoration shop owner, who was a friend of mine, called me and told me that he had a set of Buick Rally Wheels, 15 inch, with new tires on them. I jumped at the chance to upgrade. As we all know, the difference in stance from the 14-inch wheels to the 15-inch is like the difference between night and day. It's the contrast between your

Mom's Skylark and that beast they call the GS! (Well, that and the motor...) The wheels were in almost mint condition. The tires were brand new and still had the little rubber "fingers" (technically called "spew vents") on them indicating that they had never been in contact with the road. I was excited as my friend and I immediately swapped them onto the car. They looked outstanding!

The next week I enjoyed driving the car to work almost daily. On Friday afternoon around 4:30, I headed to my wife's workplace with the intention of picking her up and taking her out to dinner. It was a warm, sunny day with temperatures in the low 80's. The top was down on the Sun Coupe and the radio was playing as I hit the highway.

I was cruising at about 70 when, in an instant, everything changed. One second I was cruising in the right lane, the next second the car was jerked clear across the left lane, down a small embankment and towards the guardrail! I was able to pull the car back and get it to stop mere inches from the guardrail-so close that I could not open the door. After the dust cleared, I drove ahead and to the right so I could get out of the car. As I exited, I could see that the right front tire had literally peeled off. I went to the road and picked up sections of tire that were about two feet long and the full width of the tire. There was no blowout; the tire tread had just peeled like a banana.

I called a friend with my cell phone and he came and helped me put the spare on. We then headed back to the shop where I had purchased the wheels and tires. Turns out the tires that looked brand new were all "dry rotted" with many cracks between the treads. With the weight of the car on the tires, the cracks became more prominent. When I accelerated to 70 on the highway the tire literally peeled apart. Talking it over with the fellow who sold me the wheels and tires, he calculated that the "new" tires had been on the rims for 20 years while the car was worked on with no thoughts it would ever take that long. I was lucky in that there was no major damage to my car, and when I crossed the highway there was no other traffic so no one was injured, or-God forbid-worse!

All tires since 2000 have a date on the sidewall that indicates when they were manufactured. If you want to determine the date, locate the Department of Transportation code on the tire which begins with DOT and ends with the week and year of manufacture. Check the dates on the tires of all your cars for the safety of everyone, and remember: New is not always new!

"THESE TIRES ARE BRAND NEW, NO MILES ON THEM... 20 YEARS AGO."

Tired of this Trip
By Bill Lyon
Hooksett, New Hampshire

In 1982, a group of us were driving our street rods back home from the NSRA Nationals in Columbus, Ohio. My wife and I, with myself driving, were in the lead in our '32 Ford two-door sedan. As we exited a toll booth in the far-right lane, I checked my rear-view mirror to see if the other rods were close by. I didn't see any, so I slowed to allow them to catch up. A tractor-trailer was in the center lane and wanted to move to the right. As I was now in no hurry, I flashed my headlights to signal him that it was okay to change lanes. Just after entering my lane, the outside left rear tire of the trailer exploded and sent a recap back toward me. I had essentially no time to react. The recap was stretched horizontally when it hit the car. It destroyed my radiator splash apron. The left end of the recap wrapped around my left front fender and buckled it and the other end went under the right tire, which lifted up the car. It then came out from under the car and somehow slapped the cowl right in front of the windshield, leaving heavy marks, before again going under the car. We were bouncing around violently as I continued over the debris. In my rear-view mirror, I saw the recap twisting down the road like a snake into the path of a white Trans-Am, hitting it on the driver's side door. We both pulled over to check damage and while I looked at the '32 and tried to calm down, the Trans-Am drove off. Inspecting underneath my car, I saw that the oil filter and pan had been struck and dented and the filter was tilted backwards a little. Both pan and filter were dripping oil. My drive shaft, which is painted silver, had black spirals down its length. Meanwhile, the truck had also pulled over. I jumped back into my '32 and parked in front of the 18-wheeler to prevent him from leaving and to protect my car from possibly being rear ended.

My wife alerted our friends by CB radio so they all stopped to check on us. The truck driver was cool and very concerned for our welfare. He stated that when the tire blew, it sounded like an M-80 assault rifle going off and all he saw was a big smoke screen behind the truck and was afraid that our car had turned upside down. We exchanged insurance info and my wife and I continued home shaky and astounded that we were not hurt. It could have been way worse.

I got two repair estimates from body shop people that I know. Both were contacted by the insurance company about the high estimates and both said that because the car is an expensive street rod, it would cost that much.

A couple of weeks later an adjuster called and asked to see the car. I gave him my address and explained that the car was in my garage and could be seen any evening after 6:00 PM, as I had to work every day. He did not like that answer and told me to leave it outside so that he could drive by at his convenience and look it over. We then had a one-way conversation about the fact that I am the injured party and that maybe we wouldn't be in this conversation if the insured had properly maintained his trucks. I stated to him that maybe, instead of giving me a hard time, he would be more comfortable talking with my attorney. He suddenly agreed and we set a date for a get-together. I decided to push him some more and told him not to be late, because if he were, I would not be there.

He was on time. I met him (and another fellow he had with him) in my driveway. He looked around for the car and I again had to explain that the car was in the garage. He then told his partner to get the "crash book." By this time, I had had enough of his ignorance and asked him if he had read any of the reports of the vehicles in the accident and in what chapter of his book there was a listing of 1932 Ford Tudor Sedan parts. We entered the garage and his facial expression was priceless as he walked around the car and I heard him say to his buddy "How do we figure this?" I jumped into the conversation and suggested that maybe he should use the estimates that I supplied. I received the check about two weeks later and of course fixed the car myself.

A Great Save at The Great Race
By Art Stultz
Colchester, Vermont

The advertisement for the annual The Great Race describes it as a time/speed endurance rally for vintage cars. This race has been held each year since 1983. It encompasses driving routes in many of the United States as well as Mexico and Canada. The vehicles in the race must be 1972 model year or older and must pass strict technical requirements to be eligible to compete.

The 2018 version of The Great Race ran for nine days starting on June 23 in Buffalo, New York and ending in Halifax, Nova Scotia, Canada on July 1.

The route included passing through northern New Hampshire close to the popular 2688-foot-high Mount Washington.

The great mountain is well known, not only for its over 6,000-foot altitude and its cold and harsh winters, but for its Auto Road that twists and winds up a demanding twelve percent grade and challenges any car to climb up and safely return to the base.

Organizers obviously did not dare make it compulsory that all cars had to climb the Auto Road, but instead made it optional with the caveat that all cars that chose to make the ascent, pass a rigid test which included the performance capabilities of the car's engine, transmission and brakes. The drivers whose cars passed the tests were promised spectacular views of New Hampshire from far above the valley.

Despite the testing, the brakes on a 1955 Buick Estate wagon driven by Carl Schneider and Jack Juratovic failed shortly after the start of its descent. You can imagine the fear within the driver and passenger as the driver had to make a quick assessment as to how to safely slow and stop the huge station wagon. His choice was to steer his car to the edge of the road where he slowed by scraping the left side of the car on the mountain itself. The occupants of the car ahead, a beautifully restored 1964 ½ Ford Mustang owned by Scott and Mallory Henderson, noticed the potential disaster. They slowed their car and maneuvered it so as to make contact with the Buick and slow it even further. Both cars were thus able to stop safely, albeit, with plenty of sheet-metal damage. They were able to get down the mountain without further duress, get their cars repaired and continue with the race.

Nuts to You
By Lyle Smith
Troutville, Virginia

Buddy McWilliams, the NSRA Virginia State Safety Inspector and I were doing a walk around the spacious grounds of the NSRA Street Rod Nationals in Louisville, Kentucky. Buddy asked a couple if they would like to have their Ford Model A sedan inspected. The fellow declined but said that he had a gas leak and would appreciate it if we could look the car over and perhaps find the leak. Even though it wasn't a full inspection, it was part of our efforts to make street rodding a safe hobby so we looked for the leak. We found a leaking fitting on the fuel pump that did not have Teflon sealant tape on it. Buddy went back to his car and got some tape.

While we were making the repairs, the lady told me to go ahead and do the safety inspection after all. During the procedure I went under the front and noticed the u-bolt nuts holding the rack and pinion steering unit were loose. With just my fingers I pulled off three of the four nuts that were holding the rack on. I stood up and told the owner to hold out his hand and when he did, I gave him the nuts. He asked me what they were for and he was pretty surprised to learn where they came from. I installed them back in their place and properly tightened them. We continued to give the car a thorough going over. When we had completed the inspection, he told us that he had rushed the builder to finish the car so they could attend the Nationals. As it turned out, it was done only the day before the show. They were both very happy that we did the safety inspection.

This is a very good example of the need to allow plenty of time to complete the build so the car can be inspected and road tested long before going off to a car show or another event with it. Check that street rod over thoroughly "from soup to nuts."

Shuffle Off to Buffalo; Limp Back Home
By Art Stultz
Colchester, Vermont

In 1967 I spotted a pretty decrepit '55 Chevy 150 without an engine in the back of a used car lot in Easton Maryland. I enquired about it and found it was for sale. The asking price was $25! As a third-year auto mechanics teacher in a nearby town I had very little money, but I let the fellow I talked to know I would like to purchase it on my next payday. The salesman was so anxious to "get rid of it" he gave it to me!

I got the car running with an engine and transmission that my students gave me. I drove it in Maryland and then in Vermont when we moved there in 1970. In 1991 I retired from teaching, and did a complete rebuild and lots of modification to it in the next seven years. Two features of this car were a 5 Liter Ford engine and a "tilt front end." The grille, bumper, fenders and hood went forward as a unit with electric motors and even tilted up about 55 degrees to reveal the engine.

My wife and I excitedly (at least I was excited!) went to the big Classic Chevy International Car Show in Buffalo, New York in the summer of 1998. Classic '55,'56 and '57 cars were all lined up headed the same way with a parking space between them and remained like this throughout the weekend.

The Saturday morning of the show the "Ford Engine in a Chevy" was pretty much well accepted by the many spectators and Classic Car owners coming by. Many of the onlookers asked me about the "tilt front end," so I demonstrated how it operated with the remote-control transmitter, power seat base from a Mercury and twin generators that were rewired to be motors. It came time to start the engine and...no go! It wouldn't turn over. Probably the battery was down due to the high current draw of the tilt front end motors. I didn't realize the transmission was in reverse and the neutral safety switch was doing its job. I got the help of another car guy to do some troubleshooting. I explained to him that he could use a screwdriver to connect two posts on the Ford cranking relay together while I sat behind the wheel. We were both on the right side of the car where the relay was. I headed around the back of the car, but without waiting for me to get to the driver's seat he made the connection and the engine started up! I had left the key in the "on" position! The car was now headed in reverse across the parking lot right toward other cars and people in lawn chairs. The door handles were "shaved" and the passenger window was closed, so I had to run around to the driver's side to get inside and stop the car. As I cleared the rear of the car trying not to get run over by my seven years of hard labor, a fellow nearby opened the door and jumped in and turned the key to the off position. This was all well and good except the driver's door being open jacked right into the left side front fender of a beautiful '57 Cameo pickup! Quite a bit of damage to both vehicles.

No one was hurt. I apologized to the owner of the Cameo and explained how the accident happened. He was quite understanding considering the disaster that it was. I did the insurance thing with him. A couple of other Chevy guys volunteered to help me with getting my car drivable.

The drivers' door was pretty much crunched and bent and the two door hinges were beyond hope. There was a swap meet on the grounds, with all classic Chevy parts, so we found some used hinges and put them on. This allowed the door to close enough so I could rope it shut for the drive home.

If this wasn't enough trouble for one weekend, when we got ready to go out to dinner that Saturday night, I had no turn signals. I diagnosed that situation the next morning and felt I needed a new directional switch. I asked a local person for directions to an auto parts store, found it and bought a new switch. I didn't have a steering wheel puller with me so elected to just wire the new unit into the wires that I could get to toward the bottom of the steering column. I did this with much difficulty but that gave me directional signals and brake lights. I then instructed my wife how to operate the switch on the way home to indicate right or left turn. In due time I put the switch in properly, got a new door skin put on the driver's door and repainted it. Whew!

Pumpin' Right Along...Maybe!

Jim Knaack
Clarkesville, Maryland

Sometime in the mid '80s I left my home in Maryland in my 1928 Model A Ford street rod on my way to Milwaukee, Wisconsin to visit my parents. From there I planned to go to a car show in Minnesota.

In the center of Chicago, of all places, I pulled off the road as my engine started acting up. I was stopped no more than five minutes when two guys in beat up street rods pulled up behind me. They offered to help so I told them that it seemed like the engine was starving for fuel and I suspected that the mechanical fuel pump had failed. One of them rummaged through his jam-packed trunk and came up with not one, but two electric fuel pumps! I selected one and figured I could "hot wire" it to the alternator output wire as it was the most convenient place in this emergency. I could later install a "kill switch", when I could find one, since the alternator output is always hot. I rigged up rubber hoses and hose clamps that this guy had, and plumbed it into the fuel delivery line. I tied the pump on to the engine at a convenient spot with cable ties and the whole affair worked pretty well. The two guys (who didn't charge me a cent) followed me for a little while. I continued on to Milwaukee with just a bit of vibration from the pump as it wasn't secured very well.

I attended the show in Minnesota and then went back to my parents' home. There I got a new mechanical fuel pump at a local auto parts store and put it on. I kept the electric pump with me in case of another emergency and actually used it three more times!

Most street rodders will bring a tool box with them on trips. It's also a good idea to carry a flashlight, emergency flares and duplicate parts like distributors, starters carburetors, fan belts, etc. No telling when you or someone you come across along the way will need them. Let's be "Brothers of the Wrench" and help each other out!

Takin' it for Granted
Bob Powell
Belleview, Florida

Since I have been a street rodder, I have always been very conscious about safety. To minimize the chances of accidents, I build all my street rods with safety in mind and think about every phase of driving them such as steering, braking and handling. They may be street rods, but they are vehicles on the road and I owe it to myself, my family and others to keep everyone safe from harm.

When I lived just outside of Milwaukee, Wisconsin, I found out about a '32 Ford for sale in mid-state Ohio. I talked it over with the owner on the phone and feeling pretty good about the possibility of buying it from him, I took a bus to Ohio to get it. It was a trip of about 350 miles. My plan was to buy the car and immediately drive it home. When I got there, met the owner and saw the car, I looked it over fairly well and proclaimed to myself that this was a really top notch, good-looking machine. We made the transaction and I started home with it.

It was an older build and the odometer showed only about 600 miles, which was perhaps a "mixed blessing." It was obviously a "Trailer Queen!" Surely there wasn't much wear and tear on it with only 600 miles on the odometer, but was it road tested enough to find out about all the "kinks" in it? Probably it would be okay.

Wrong! There were many things that were far from satisfactory. I will only speak to the safety problems here.

While driving in a steady rain, without wipers, which was bad enough, I soon discovered that the speedometer accuracy was way off and the fuel level gauge read a half tank while full, as well as empty! Those two defects made it hard to tell just when to fill up with fuel. I stopped along the way and made a cursory check of the suspension to find that several of the castellated nuts were quite loose but at least they had cotter pins in them. I tightened them by hand, which was all I could do without many tools with me. I didn't notice until I got home, but I found that the big nut which held the Pitman arm to the Pitman shaft had loosened and was on a mere two revolutions! Unfortunately, I had assumed that all the looseness in the steering was from unseated tie rod ends or something similar. I was very lucky that day, and looking back I should have trailered it home and inspected it thoroughly later when I had plenty of time.

I did manage to get home safely and eventually went over the whole car checking everything I could think of. There were quite a few improvements needed.

Thinking I had found everything, I took the car for a test drive down the road and got on the brakes pretty hard. The car pulled violently to the right sending me toward the ditch! The lower caliper bolt had backed out causing the caliper to swing out and jam the wheel. I replaced all the caliper bolts with Grade 8 and put lock washers and

Loctite on all of the caliper mounting bolts, which solved the problem. Since these custom aftermarket calipers are held in place only by clamp load, they should probably be safety wired.

This was an eye-opening lesson on checking a car over before driving it extensively. A street rodder might find a nice car in an ad in a newspaper, Hemming's Motor News or StreetScene magazine and would like to purchase it even though it is many miles away. It is very tempting to take the owner's word for it that the car is in safe condition. Perhaps a better policy is for the new owner to "flatbed" it home so he can take his time and check it over when he has it in his garage.

Yankee Ingenuity...in Pennsylvania
By Jim Ricker
Colchester, Vermont

In the fall of 2000, I went to two big car shows in Hershey and Carlisle Pennsylvania. They have big "Flea Markets" at both places. I have lots of automotive related stuff to sell, so I go to them almost yearly. The shows are fairly close to each other and are on consecutive weekends, so it makes it convenient to take in both events.

I drove my 1985 Ford 350 TransStar camper which was powered by a big 460 cubic inch engine in front of a Doug Nash overdrive transmission. It pulled my flatbed loaded with many things for sale. The shows offer rental spaces for anyone that wants to sell parts or cars to the hundreds of attendees. I have had success buying and selling cars and other related items over the years.

This year there was also an auction of cars in the city of Carlisle as well. The auction organizers planned it for earlier in the week in hopes of drawing people like myself who were going to the big show. A popular race car builder had passed away and his shop and many of his cars were up for auction. One of the featured cars was a 1933 Ford five window coupe. This street rod had a 351 Ford Windsor engine, '39 Ford transmission and the very desirable Ford nine-inch rear end. It also had a very well-built race car suspension. I went to the auction, liked the car and bought it.

Since the plan was to take in the auction and then participate at Carlisle and Hershey, I took the '33 on the flatbed to the shows.

I kept the car on the trailer throughout the Carlisle show while I sold quite a few car parts. Some participants spotted the '33 and showed a great interest in buying it, but I told them I had just purchased it and did not intend to sell it. This would prove to be a good decision later on the trip back home.

On to Hershey the next weekend. Hershey is a different show from Carlisle in that they don't cater to hot rods as much. An official who spotted the '33 said "Get it off the field or at least cover it up." I covered it up with the car cover that came with the car.

I left Hershey on that Saturday around mid-afternoon. I got on Interstate 81 which eventually became Interstate 78 and headed for New Jersey and then home. I drove for about twenty miles and when I came upon Exit 3 of 78 the engine of the camper simply stopped running! This was a big, heavy rig to deal with considering the trailer was loaded with the car and lots of heavy items including a Triumph T120C motorcycle, a big Binks air compressor, car parts, tools and many other Flea Market items. I nursed the rig to the side of the road the best I could and parked it. There was a truck stop at this exit so I walked over to it and asked a fellow who worked there where the nearest Ford dealer was located. Since the engine was a Ford, a Ford dealer with a service department would be best for diagnosis and repair. The fellow said it was 18 miles away but he was sure it closed at 3:00 PM on Saturdays. It was past that time already.

I was getting concerned as to how I was going to get a wrecker to tow this great heavy assemblage to a Ford dealer or repair shop. There wouldn't be anyone working on that Sunday, so probably I would have to spend a couple of days and nights waiting for it to be fixed.

Okay. I thought, "I'll try to fix it myself." I tested for spark. That was good so perhaps the problem was lack of fuel. The engine had an electric fuel pump with a relay. Maybe the relay had gone bad. I had a spare relay with me as the former owner had suspicions that it might be bad. Replacing the relay with the spare didn't cure the problem. I had electricity at the pump but got no action from it. What to do now...

Aha! The '33 street rod had an electric fuel pump! The pump was in the trunk and fairly accessible; not hidden down under the body near the fuel tank. I removed it from the car and hot-wired it to test it. It worked! Now to put it on the camper engine. This was a matter of disconnecting the fuel lines and plumbing it in in place of the bad pump. When I hot-wired it again and cranked the engine it started up. Good deal. I had some cable ties with me so I fastened the pump to something nearby. It would have been a time-consuming job to do the wiring correctly so the pump would operate with the ignition switch. I simply fastened it securely to the positive post of the battery. This would eliminate anything defective that was normally wired between the battery and the pump.

The engine ran but skipped a bit. The carburetor was of the type that had a fuel return line which helps keep the fuel circulating and preventing vapor lock. I pinched it closed with vice grips thinking that might be advantageous. It did seem to help so off I went. After seven hours of driving I arrived home in northern Vermont.

We New Englanders are known as "Yankees" and the work I did surely could be described by another popular term: "Yankee Ingenuity."

Air Conditioning...at Any Cost!
By Walt Fuller
Newnan, Georgia

I had been looking for a nice '56 Chevy for some time and finally found one. I bought it in July of 2017. It was a blue and white 210 model.

In April of 2018 my wife and I were headed with the car to the Southeast Street Rod Nationals in Tampa, Florida when we stopped at an Arby's restaurant in Lake City for lunch.

The air conditioning hadn't been working as well as it should have been so this was a good time to check it out. I left the engine running with my wife in the passenger's seat and went out and lifted the hood. The system was a combination heating and A/C unit so I thought I may as well shut off the valve in the hot water delivery tube just in case that had anything to do with the A/C system being sub-par.

I then went into the car by way of the passenger side and reached out to operate the levers on the A/C control panel and inadvertently brushed against the transmission shifter, knocking it into reverse. The car of course started moving rearward. I hollered for my wife to turn the ignition off but she was thinking she ought to put her foot on the brake pedal. In the excitement to get the car stopped the shift lever was pushed forward

into drive. My wife tried to push on the brake pedal and apparently caught the gas pedal instead, so the car shot forward with the engine running at maybe 4000 RPM. It then crashed with full force into a power pole which was just ahead of the car!

Everything between the fenders was extensively damaged. That included the bumper, radiator, hood, A/C condenser and grille. The fenders had just mild damage. We weren't injured except for bruising of my ribs.

There was a deputy sheriff nearby who saw it all. He checked on our well-being and then said that since the incident was just outside the city limits it fell within his jurisdiction so he was obligated to "call it in." There was $11,000 damage to the car and to "add insult to injury" I was assessed a $167 fine for negligent driving!

The car was towed to a wrecking yard where it stayed until I could get back with a flatbed and take it home. We rented a car locally to get home ourselves.

Not a good day to say the least. Looking back at the whole affair it would have been very simple to avoid it. Just don't touch the shift lever. Putting on the emergency brake would probably have avoided the incident entirely or at least lessened the damage. This was quite an important lesson regarding just how much damage can be done by not making the right decision in a split second. What was done in that brief amount of time resulted in many months of repair time and the spending of thousands of dollars as well.

Island Hopping
By Ken Bessette
Williston, Vermont

I have a 1953 Studebaker Commander that I acquired in 2014. The previous two owners had done extensive work to it, including removing the rear doors and an equal section of 34 inches from the frame, which thus created a "Shorty." I did lots of work to it in the next couple of years to make it safe and easy to drive.

I inspected many things on the car including the brake lines from front to rear and decided that they looked pretty good as is.

In 2016 I took the car to the NSRA Northeast Street Rod Nationals in Essex Junction, Vermont. Thursday was the "setup day" before the three-day weekend show. I had some work to do to help with the preparations for the show so spent some time there and then headed out in the late afternoon to pick up my wife Gayle at the host hotel as she was helping with the car registrations.

I was cruising along the busy four-lane thoroughfare as fast as 45 miles per hour. Cars ahead were stopping for a traffic light so I casually put my foot on the brake pedal and it went right to the floor! What to do next! There were only a few choices including swerving hard to the right to avoid the street rod that was stopping in front of me. It flashed through my head that I had recently spent $1600 to get the twin grille sections of the car re-chromed so I thought this choice was a good one. At this intersection there was a small "island" that separated two lanes coming from the right out of Camp Johnson, a military base. I first jumped the curb to my right, then up onto the curb of the island, then another curb onto the side of the road beyond the stopped traffic. Luckily there was no vehicle exiting Camp Johnson as I surely would have lost my re-chromed grille amongst other car parts and pieces. I managed to coast to a stop on the side of the roadway.

I then called my son with my cell phone. He got my pickup truck and trailer and we loaded the car on the trailer and took it home. Needless to say, the car missed the show that weekend.

Upon further inspection of the brake lines, sure enough I had missed seeing a section of line that was rusted through and gave way under the stopping pressure.

I later pulled out the engine for much better access and replaced all the brake lines. Looking back, I had inspected the lines but obviously not well enough. I learned a good lesson about inspecting someone else's work thoroughly.

Bearings, Crashes and Burning Oil
By Art Stultz
Colchester, Vermont

During the summer of 1959, I purchased a cool '54 Ford two-door sedan with a 239 CI V8 "Y block" engine. With my limited knowledge of building custom cars at that time, I still was able to "nose and deck" it, lower it with a lowering block kit, add a single glass pack muffler and a "Bermuda buggy bell." I always liked the look of the '48 Mercury grille so found a junk one and installed it in place of the original grille. I enjoyed learning the basics of how to tune and maintain it while I attended classes the next academic year in graduate school at the University of Vermont.

The following June the Army was after me, so I enlisted in the U.S. Air Force to get my military obligation over with and out of the way. While stationed in San Antonio, Texas, at Brooks Air Force Base with the Apollo Space Program, I discovered a night course in Auto Mechanics at a local junior college. I enrolled in the program and ultimately received the Associate of Arts Degree in Industrial Education, majoring in Auto Mechanics. The first course I took that school year, 1960-'61, was Engines.

I was anxious to go on military leave the following August and go back home for a couple of weeks, and get my '54 and bring it back to the air base. My parents thankfully agreed to store it in a spare building there on their rural property which at one time was a farm. The big nationwide push for installing and using seat belts had started about then, so I bought a set of seat belts at the air base at a bargain price. At home I installed them on the car for the trip back. I thus started out toward the end of my month-long leave headed out to Texas with expectations of a good, but long, drive. I suspected there was a problem with the engine as it was making some rattling noises now and then, which led me to suspect that the main and/or rod bearings were not what they should be. I noticed it was also going through oil at a pretty good clip as I checked the level on the dipstick often. I added oil now and then from a gallon oil container I purchased along the way. Somewhere in Ohio, the oil "idiot light" started coming on more often than it had been, so I figured I'd better stop at a repair shop and get it checked out. I came upon a good looking, big shop in Dayton, Ohio, and they installed an oil pressure test gauge and notified me that probably the bearings were wearing out and perhaps the crankshaft journals were scored. I knew what that was all about as I had just studied it at the college in San Antonio. They gave me a price and I told them to go ahead with the job. I had to wait the three days that it took, so luckily Wright-Patterson Air Force Base was right nearby and I could stay there overnight free with my military ID. I didn't have the money for the repair so had to arrange for my dad to send me some money from his bank in Vermont to a nearby bank there in Dayton. I managed to get through the red tape of the transaction, although it was a hassle.

Back on the road again. Troubles over. Or so I thought! The repair shop guy said the price would cover the bearings and journals but not the piston rings so I had to keep adding oil, but that was not too big of a deal.

Somewhere in Kentucky I was going right along at a good clip when I came over a hill and a half dozen cars were stopped dead still in the road ahead! They were waiting for a tractor that was at the head of the line to make a left turn when traffic cleared. I slammed on the brakes but to no avail, as I rammed into the rear of a new Chevy ahead of me and from there got pitched off to the right and into the ditch. As it turned out, I was the last of several of the cars that had struck each other in a "chain reaction." Traffic slowed down and stopped, the State Police arrived and thankfully no one was hurt. They called a tow truck for me. The driver had his own small backwoods shop so towed me there. The left front fender was pretty well jammed up as was the grille with my nice '48 Merc modification. The radiator was done for. He spent quite a while finding a used radiator for me and putting it in and simply getting the car roadworthy and safe to drive. We agreed that he would not do much to fix the fender and grille. The left headlight was cocked over and aimed toward the ground but that was OK. The bent-up hood stayed latched for the rest of the trip. My Dad had convinced me to get insurance before I left so I had done that, thank goodness.

I continued the long trip from there through the mid-South and got pretty tired. One night I tried to drive all night but just couldn't do it. I found a sleazy three- level hotel in a small Arkansas town about five in the morning and rented a room for very little money. I slept until about 10 AM. There was no bathroom in each room so customers had to use a general restroom for each floor. I had to go down one level as the toilet on my floor was all "stuffed up" and stinky. I got going back on the road after adding a little oil to my '54. One morning I stopped at a little southern restaurant and ordered some eggs and the waitress said "Y'all want greets?" I had to ask her twice what in the hang she meant and she finally slowly said, "Do...ya...all...want grits?" She had to explain what grits were as I had never heard of them. I said sure. They were OK.

Finally limped into the Air Force base. I had fortunately allowed a few extra days' travel to account for the eventuality that I actually had, so was not late going back to work. I approached the Air Police shack at the entrance to the base and the AP on duty stopped me because I didn't yet have the required windshield sticker for base residents. He shook his head when he saw the front of the car and said, "Your car will be classified as 'unsightly' and have to be fixed." (Hmmm...thanks for the observation!) I eventually found a replacement left front fender and a hood, discarded the banged-up Mercury grille and got the car looking pretty good. I sold it to another GI and bought a '49 Mercury two-door sedan with the flathead 256 CI engine. It had its troubles but we'll save that for another story.

Whoa Effie!
By Dan Sargent
Swanton, Vermont

In the fall of 2004, my Dad saw a 1948 Ford F-1 pickup truck street rod parked at a repair shop. He noted a for-sale sign taped to the windshield and thought he just "had to have it" so he went back and bought it the next day. Five miles into the drive home he realized that it was not roadworthy. The truck did not steer well and it shook and vibrated horribly going down the road. Among other things, the brakes didn't work well, the rear axle housing and differential were not installed properly as the pinion angle was way off, and the steering box and column bearings were worn out which caused a lot of play in the steering and a lack of good control.

Dad took the truck to a local street rod shop that had recently opened for business. The owner had previously specialized in motorcycle repairs but was branching out into the world of street rods. Dad paid him a hefty down payment in hopes he would have a drivable street rod to enjoy. Weeks turned into months and work progressed slowly. Dad would stop by to check on the progress and would often work on some things himself, such as installing a new disc brake conversion kit and cleaning engine parts. By the following spring, the truck was deemed done. After paying another hefty sum, Dad drove her home with new front disc brakes, a properly aligned driveline, and a rebuilt engine put together from parts from two similar engines.

Everything went well for a couple of years, then in 2006 Dad took the truck to a small car show, the annual Bean Chevrolet Car Show in Northfield, Vermont. My brother Tim and I also attended. Tim drove his 1948 Pontiac and I rode with him. After registering his truck, Dad proceeded to drive through the rows of parked show cars to an open parking spot. As he was turning left to park the truck, the engine raced uncontrollably and the rear wheels began to spin, spewing grass and mud into the air. We heard the commotion and looked over to see Dad with a panicked look on his face, clutching the steering wheel and apparently jamming on the brakes. With brake lights on and the engine revving, the truck spun sideways, just missing a parked show car as Dad reached down and turned the key to shut the engine off. We ran over to ask him what happened.

He said he had no idea. We popped the hood and inspected the throttle linkage, the carburetor throttle plates, etc. We could not find any binding or such that would cause the engine to race like that. Our assumption was that Dad had possibly been pressing on the accelerator and brake simultaneously without realizing it. Perhaps it was a vacuum leak on or around the carburetor which caused the engine to race.

When the show ended, we waited for most of the other cars to leave before we did. I rode in the truck with Dad and Tim followed in his car. The first part of the ride went smoothly and Dad and I decided to stop at a local auto parts store to look for some rubber vacuum port plugs for the carburetor. As Dad was turning left into the parking lot of the store, the engine began to race again. Dad turned the key off and coasted into a parking spot. We purchased a package of vacuum port plugs and replaced all the old dry rotted ones right then and there. The engine started right up and we revved it a few times by hand and still found no evidence of a sticking throttle.

Tim asked Dad to get in the truck and work the throttle pedal. He did. Still no evidence of binding. It then dawned on us that each time the engine raced Dad had been making a left-hand turn. We told Dad to turn the steering wheel slowly to the left while holding his foot on the brake pedal. As he did so, the engine RPMs increased. Finally, we found the culprit was a combination of a few things. The engine had been re-installed with old worn out engine mounts. The brakes had been updated with a new power disc brake kit with the vacuum assist unit and master cylinder mounted on the old and weak firewall. The third issue was the new throttle cable had been installed with no slack in the cable housing between the firewall and the carburetor. So, when Dad applied the brakes, the old unbraced fire wall flexed which created tension on the throttle cable, pulling the throttle open slightly which in turn caused the engine to flex on the worn motor mounts which pulled on the throttle cable even more, all without pressing on the gas pedal!

My Dad with "Effie."

After installing new motor mounts, properly bracing the firewall around the power brake booster, and installing a new longer throttle cable with adequate slack, the problem was fixed.

Over the next couple of years, Tim and I worked on the truck with Dad. We re-did most of the work that had been done at the street rod shop. This just goes to show that just because someone says they know how to build a street rod, it doesn't always mean they can do it properly and safely.

Dad, Tim and I would have "family hot rod nights" where we would pick an issue to work on, drink a few beers, and bust a few knuckles. We affectionately called the truck the 'F'ing Ford' and eventually nicknamed her "Effie" for short. Our Dad has since passed away and Tim and I are now Effie's caretakers. We still work on her to keep her on the road, all the while still drinking a couple of beers and busting a few knuckles.

The Invincible Wall
By Roger Kendrew
Milton, Vermont

I have been a street rodder for many years and have owned lots of rods and customs and been to many shows. I started off with motorcycles however, way back in the mid '60s.

I had just finished my junior year of high school in the summer of 1966 and worked at an ARCO gas station on North Avenue in Burlington, Vermont.

North Avenue is a long, busy street which stretches out to the north of the city. About half way out, is a big park called Ethan Allen Park, named after the Revolutionary War hero and one of the founders of the state of Vermont. There is a high stone wall around the park with an entry and exit gate. Lots of local youth like myself would drive their early street rods and fathers' family cars like crazy through the park with its narrow, hilly, twisty roads at dizzying speeds showing off to their buddies. I never heard any horror stories of crashes at the park but there probably were some.

My story involves the wall of the park. The ARCO station where I worked was right near the wall. A young fellow who was a frequent customer at the station had a '65 Yamaha Big Bear bike with a 250 c.c. engine. I would borrow his bike now and then and "go for a spin" in the neighborhood. At age 16, I was young and of course felt "invincible."

I was buzzing recklessly around the gas pumps on the Fourth of July on the borrowed motorcycle when the throttle cable stuck somehow. The bike carried me out of control; first onto North Avenue where I hit a curbing and then back toward, and head first into the wall!

In those days no indestructible youth of sixteen would wear a helmet so I was pretty banged up and "out of it" for a while, but I didn't lose consciousness. There were lots of people in the busy area. The person that I remember who seemed to be the one who called the ambulance, was a good looking "well endowed" shall we say, young lady. (Funny how we remember the important things!)

When I eventually talked things over with the owner of the bike, he insisted that there was nothing wrong with the controls when he loaned it to me. I insisted that there was something about the throttle cable, such as being overtightened, that caused it to hang up and not release as it should. He was very reticent to talk with me about it, which is understandable. I never knew just what he did with the totaled motorcycle after he got it back. Probably he simply got reimbursed by his insurance company and let it go at that.

I spent eleven hours in Intensive Care at the big hospital in town with a broken hip and fractured skull. I was six months on crutches but was able to go back to school as a senior when it re-opened the following September. By then my buddies had heard about the incident and kidded me that with that stunt I was now fully accepted into the "Hells Angels" Motorcycle Club! The high school had a Trade School which included an Automotive Technology course that I was enrolled in. This not only gave me the training and experience for a vocation as an auto mechanic, but fed my interest in cars and street rods as well. I belonged to a local street rod club for a long time. I still limp a little from the big crash so many years ago. I don't feel quite as invincible these days as I did then!

The Price of a Free '55 Chevy
By George Lucia
North Ferrisburgh, Vermont

There are eighteen members in my car club, The Champlain Valley Street Rodders. We have been a club since 1981 so of course many members have joined and many have resigned over the years.

At a club meeting on June 19, 2018, a fellow in the club, Gary Bruening, announced that he had retired from his long-time job and he and his wife were going to move to Rhode Island very soon. Gary had lots of accumulated parts and pieces of street rods, like we all do, and told us that he was planning to take his '38 Chrysler and '67 Pontiac with him. He would make several trips south with his trailer. Mostly in jest, I spoke right up and said "I'll take your '55 Chevy off your hands if you like!" With no hesitation he replied: "I tell you what George...if you come and get it in the next few days, it's yours."

Wow! What a shock! I have a '48 Chevy pickup with a 348 CI engine that is far from done and hasn't been on the road at all, but as the saying goes, "We street rodders can't have enough toys." I immediately said "sure...I'll take it."

The '55 hadn't been on the road since about 1980. Gary had purchased it in 1999 from another guy in the club. It was pretty shabby and had no engine or transmission, but it was a much sought-after Bel Air and had lots of possibilities.

The best day I had to go to his place to get it was on Saturday June 30. Early that day, I hitched up my big two-wheeled trailer to my every-day pickup truck and drove north from my home in North Ferrisburgh to Cambridge; some 45 miles away. I got to Gary's home about 7 a.m. Gary had some other items he was going to give me as well as the '55, such as a good frame to replace the bad one on the car, and some engines and transmissions. The plan was to put the '55 on his trailer and we would put the other things on mine.

Gary's garage had a sloping driveway such that we were able to back his trailer up to a point where the '55 could be pushed out the door and onto the trailer. The car didn't have a center link or tie rods, so we had to kick the wheels as we pushed it so it would track onto the trailer. We managed to load my trailer with all the other items.

In about 40 minutes, we headed to my place with me leading. Everything was fine for 35 miles until I felt some awful "carrying on" with the trailer I was pulling. I pulled off to the side of the road and discovered that the right-side tire had almost totally shredded. We decided to pull the trailer a few miles to the nearby high school which had a huge parking lot. Being a Saturday and summer vacation, there probably would be no other vehicles in the lot and few, if any, people around. I knew that if the cops saw us driving with a flopping trailer tire making black squiggle marks on the road, they would surely get after us, so we intentionally took the most back roads we could. There was a parade

Street Rod Horror Stories

commemorating the upcoming Fourth of July and honoring the local fire departments queuing up in the nearby town. We figured all the local police would be there and not bother us. After parking my truck and trailer at the parking lot, we went in Gary's truck to my house and unloaded the car off of the trailer.

I told Gary I had a spare wheel with a tire mounted on it, somewhere in the basement of my house, so I sent Gary on his way back home. I would have my wife, Linda, drive me back to the truck and trailer. I went down to the basement but the spare was nowhere to be found.

I called my neighbor for possible help but he wasn't home. Stanley, another fellow in the street rod club, lived nearby, but I knew he was on a cruise that the club was having that day and wouldn't be home either.

Hmmm...what to do. I decided to simply drive on the shredded tire the best I could. Linda drove me back to the truck and trailer and I drove slowly back home. When we arrived, the tire was very badly wrapped around the axle in a big tangle.

I worked for the next three hours trying to get the tire untangled. I used a "Sawzall," then tinsnips, and finally a come-along. I fastened one end of the come-along to the tire and the other end to the newly acquired '55 Chevy and eventually, with various positioning of the hook, pulled the tire off the wheel and away from the axle. What a job!

I figured that Stanley would be back by then, so I took the wheel with the bits and pieces of tire still on the rim the short distance to his place. He had indeed returned home so he helped me grind off the pieces of tire. He had a tire just the right size, as well as a tire changing machine, so we put the assembly together. I was able to go back home and put it on the trailer.

Tread good...sidewall bad!

The interesting thing was that the bad tire had just sidewall rot and damage; the tread was fine! Just goes to show you that there is more to a tire wearing out and getting old than just the tread.

Interesting fact number two was that to this day, I have never found the spare wheel and tire that was supposed to be in my basement. Perhaps I loaned it to someone. If so, I have forgotten who that was…

The whole affair was a hassle, for sure, but a small price to pay for a genuine 1955 Chevrolet Bel Air!

Encounter with a Stone Wall
By Doug Juonis and Diana Bishop
Litchfield, New Hampshire

Diana and I went to the Northeast Street Rod Nationals in Essex Junction, Vermont in September of 2006. We were doing about sixty in our '31 Ford roadster on Interstate 89 near Barre, Vermont about 30 miles east of where the show is each year. We came up on a car ahead of us and were turning into the passing lane when a car came up very quickly on our left. To avoid it we veered off to the right and flew off the highway. The Interstate had been constructed in the early '60s by blasting away rock formations that left some pretty sheer stone walls. We hit one of these walls on the passenger's side and went end over end one time which totally wrecked the car. During the flip we became completely turned around in the car so instead of facing the front windshield we were facing the rear window.

Our friends Bob and Denise LaRoache were driving in front of us and saw the accident in their rear-view mirrors. They immediately called 911 and gave them the exact location. David Bishop and John Parzych were in the car behind us. Among the four of them they got us out of the vehicle before the ambulance and police arrived.

We were taken first to the Central Vermont Hospital in Barre where they set Diana's leg without anesthesia, which was a big mistake. The doctors discovered that she needed to go to a trauma one center so they took her to Dartmouth Hitchcock Hospital in Hanover, NH for further surgeries and treatment.

Diana took most of the damage with a crushed ankle, broken leg, seven skull fractures and cracked tooth fillings. She now has a titanium eye socket with screws and a titanium rod and screws in her left leg and ankle. Any bones in her face that were not broken were fractured and all the fillings in her teeth were cracked. I had bruised ribs, a sprained wrist and ten stitches over an eyebrow.

Our car had a Gibson fiberglass top which held up pretty well. Both doors stayed closed.

One of our friends stayed with the car and arranged for a ramp wrecker truck to take the car to Bob's Sunoco in Montpelier. The next day it was taken to Nashua, NH to await the insurance agent. The car was totaled. By the looks of the mangled car, we were very glad we weren't totaled as well!

Our good friend Ed Miller, who at the time was the New Hampshire State Representative of the National Street Rod Association, took me to the hospital to visit Diana and brought me back home. She spent ten days there.

It took seven years to rebuild the car after buying it back from the insurance company. The car features a 4 ½ inch stretch in the doors, torsion bar suspension, rack and pinion steering, four-wheel disc brakes and a Boyd's independent rear end. Besides the seat belts, the safety features and build of this car is what saved our lives! We also thank all or our friends as well as the emergency personnel for all they did to get us the help we badly needed on that fateful day.

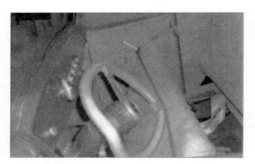

Crunching the Best
By Jim Ricker
Colchester, Vermont

What car makes the most popular street rod? The answer to this question is controversial but the 1932 Ford surely is in the running for the street rod most coveted by rodders the country over. Body designs include the three and five window coupes, two and four door sedans, Roadster, Convertible, Cabriolet, Phaeton, and Victoria. There is something appealing about the shape of the body, the curvature of the fenders, the design of the reveal moldings and especially the grille and grille shell that attracts me to this car. Thousands of other car enthusiasts have been drawn to it as well.

In 1999, I was very fortunate to be able to buy a beautiful '32 Ford Model B Roadster with the four-cylinder engine. I purchased it from Bob McGowan of Connecticut. I played bass guitar in a band with Bob back in 1975.

I replaced the "four-banger" with a '36 Ford flathead and bolted up a '39 Ford transmission behind it. I had it lookin' pretty good in 2002, so I drove it to the Lone Pine Car

Show in Colchester, Vermont in early June of that year. Lone Pine is a nice campground which is very suitable for car shows and just a few miles south of where I live.

I was going home after the show and approached a junction of three streets: the one I was traveling on, one coming at an angle from my right and one teeing in from the left. There were two northbound lanes, the left lane for a left turn and the right lane to go straight. I was in the right lane. A Toyota pickup truck was slightly ahead of me in the left lane with left turn signal blinking but suddenly the driver veered abruptly to the right and I "t boned" him in his midsection! Apparently, he decided to turn into a restaurant on my right.

The collision took out the original grille shell and insert, the radiator, both front fenders and badly crinkled the hood.

The pickup was damaged in the right rear quarter but was drivable. The driver was very apologetic; however, he did not have insurance or a driver's license! Fortunately, the "uninsured motorist" portion of my insurance took care of the costs.

The accident happened only two miles from my home so a fellow gave me a ride to my place. I got my truck and trailer and went back and picked up the car.

Later I pulled out the engine and transmission in anticipation of using that combination for a future street rod and sold the smashed up remains back to Bob McGowan.

What a shame to lose all those nicely sculpted 1932 Ford parts. At today's prices, replacements of those few parts would be (approximately):

Walker radiator	$869
Grille insert	$250
Grille shell	$260
Headlight bar	$160
Splash apron	$ 65
Fiberglass fenders	$520
Axle	$400
Total	$2524

1932 Advertisement

Before

After

In retrospect, I realize I should have kept the car as I could have rebuilt it. It was all metal with no fiberglass which is an important factor. I know that accidents happen and I know it wasn't my fault, but at the same time, I know that the mishap could have been easily prevented. No saying how many years I would have kept and enjoyed my much beloved '32 Ford roadster, had the fellow simply made the left turn he signaled for.

A Damper on My Trip
By Paul Zampieri
Barre, Vermont

It was late August 1993 when I went with my street rod club, the Vermont Street Rodders, on a trip to attend the big annual MaCungie Car Show called "Wheels of Time" in Pennsylvania. I drove my '34 Ford with an original style flathead engine (the somewhat unique 59AB) on the nine-hour trip.

We enjoyed the show the first day and as a group, we left for the motel, a Motel 6, pretty late in the evening. I was the last of the group to start up my engine when I heard a loud "clunk." I shut the engine down and got out and took a look at it. The fan belt had flipped off because the vibration damper on the front of the crankshaft had broken in half! The damper was a replacement steel damper that had four welds rather than the original. The welds wholly or partially somehow gave way.

All the other guys in my group were well on their way to the motel, so I couldn't get their attention to tell them I was unable to follow. This was before the era of cell phones, but I was able to call the motel from a phone booth there on the grounds. We had registered before attending the show so I had the information about the motel, including the phone number and the room number of one of the couples in our group. I contacted the motel but the fellow at the desk couldn't understand me. Probably English was not his first language!

Now what was I supposed to do? I was stuck at the very quiet show grounds. One and a half hours went by before Ken and Gayle Bessette sensed that I hadn't returned to the hotel, so came back to find me. By now it was 12:30 A.M.! They picked me up and we went the seven miles to the motel. We left my car there for the night.

The next morning, we got some leads from local street rodders and searched in town and around the area for some guys with Ford flathead experience who might have a damper we could replace mine with. Nothing! We went back to the show grounds and found a Marquette test equipment and welding vendor and asked if we could borrow his welder. He said we could, so Ken welded the damper back together so that it could be hammered back on securely. I put on the big center bolt, tightened it and it held very well for the trip back home, although it wobbled a bit.

Sometimes no matter how much we inspect our street rods, something like this happens that we could not have predicted beforehand, so we simply "roll with the punches" and take care of it on the road.

Loose Nuts: Preach What I Teach
By Art Stultz
Colchester, Vermont

I always work on one of my two hot rod cars during the winter months and put the other one away across town in a storage facility. This particular winter it was the '32 Ford five-window coupe's turn to get some attention.

In Vermont spring is pretty slow in coming, not to mention summer. Finally, a good day arrived when my winter work was pretty much done, and it was time to take the car out for a "spin." I live in a small town with not much of a center but there is one which consists of a library, fire station, church and various small buildings. I went the several miles to and past the town area, and was satisfied with my little road test, so turned around and headed back home. Just as I got alongside the library, I felt a severe jolt in the right rear of the car. It felt like the wheel had come off! I quickly pulled off to the right where there was a good flat area with grass and no curbing. Fortunately, this didn't happen between the library and home as the road had a few curves and twists and not any shoulder to speak of.

To my amazement, the wheel and tire assembly was rolling along its merry way to the right and ahead of me and down the roadway. It was headed for an intersection as it slowed and finally flopped onto its side nearly in the middle of the road, as a single solitary car approached the wheel and turned to avoid it. I sprinted up to the wheel, picked it up off its side and rolled it over to the edge of the road and on back to the car.

The wheel-less right rear of the car had slammed down to earth, with the vertical bracket I had built for the suspension's "trailing arm" hitting hard onto the dirt with no damage to the brake drum. The bracket was strong and well made so suffered no damage except some paint removal.

So why had the wheel come off? Simply put, the five lug nuts had come off! I figured they were scattered in back of the car somewhere. I walked back and found three out of five just as a friend of mine, Mike Bryan, came by in his car and stopped to help. I had most of the tools I needed in my car but lacked a couple, so he went back to his place and got what I needed. I always keep a bottle jack and wood blocks in the trunk for just such emergencies, so managed to get the wheel back on securely with the three lug nuts I found. Probably the other two were way back there somewhere in the grass and dirt.

I drove cautiously back to my shop and while looking for some replacement nuts, I spotted two of them nearby that looked like the ones that came off! Oh no! At some time during the winter I had put the wheel back on with three of the five nuts and apparently tightened them no more than finger tight.

I was a high school Automotive Technology instructor at the time, and one of the many safety precautions I had taught my students was to either put the wheel on all the way, torqued to specifications, or don't put it on at all. If necessary, leave a note taped to the windshield: "Tighten lug nuts" or "No brakes" or whatever message applied to the job. I had obviously not followed my own warnings!

Over the years, I have noticed that at the time of doing something, it seems like I surely will remember later on just how or why I did it. But later the mind just doesn't recall as well as it did years ago. I often find a note on a tag of an auto part, let's say, and I can't even remember writing the note much less what the note told about.

I didn't tell my students about my little nut tightening mistake…

Slippin' and a Slidin'
By Lionel "Puddy" Paris
Ipswich, Massachusetts

In August of 2016, my wife, Stephanie, and I went to a car show in Hebron, Maine. We drove our 1940 Chevy four-door sedan street rod which was towing our TrailManor camping trailer. Stephanie had just had a knee replacement the month before and the doctor told her she could go to the show as long as she was careful.

On Sunday morning we went out to breakfast with other street rodders. It was raining fairly hard when we returned to the campground to pack up and leave. In hindsight we should have stayed there at the campground until Monday!

The road surface was very slick as we cruised down Interstate 95. Although it was a Sunday, the traffic was exceptionally heavy. We were approaching the town of Kennebunk when we came over a rise in the road and were astonished to see the brake lights lit up on the cars ahead of us.

I tried to slow down but the roads were too slippery and the heavy trailer seemed to be pushing the car. Almost totally out of control, we went down between two lanes of cars. What the street rod didn't hit the trailer did! We both thought we were going to die. It seemed forever before we got our rig stopped, but it did of course, and we found ourselves up against the guardrail in the passing lane with other cars here and there around us. The trailer stayed coupled to the car but was now sideways on the highway.

When we calmed down, we tried to get out of the car, but both front doors wouldn't open from the inside as the latch mechanisms were badly jammed. Our outside door handles had been removed like lots of other customized cars. That is, the doors opened by a transmitter or hidden push button and not the normal handles. Therefore, people who came to our car to help couldn't open the doors either.

I climbed into the back seat and reached over and managed to get Stephanie's door open. Her "new" knee had snapped very loudly and her other leg had come up under the dash and had a very large lump on it. She had some major bruises as well. A motor mount had broken and the engine had been pushed into the right fender and firewall. I had a sprained ankle from my foot getting caught under the displaced pedals.

Several people came to help including a nurse. Someone called for emergency aid. An ambulance arrived and took Stephanie to the hospital to make sure her knee was okay.

The car and trailer were towed away to a garage in Kennebunk. I took a taxi to the hospital where Stephanie was and later our daughter came to take us home. We got the car and trailer back to our house a few days later. The insurance company declared the trailer a total loss so they took it away. The car was put in our garage. Stephanie could not even look at it; she told me to do whatever I wanted with it. My good friend, Les Harriman, inspected the car and said he probably could fix it, so it was taken to his garage where he repaired all the damage. We put a completely new chassis under the car and replaced the cowl with one from another similar car. We also put on a new grille and left front fender. Les straightened out other sheet metal as he had been a metal fabricator at General Electric. The car was then delivered to a place in Maine to be repainted.

The trailer was well insured and I had liability but not collision on the car, so we had to pay for the latter "out of pocket." Other cars that were banged up were covered as well. The insurance company said that the amount for covering the entire accident including the other cars was $168,000! Since the accident we have learned our lesson and have all our cars well insured.

We both had nightmares for quite a while after the accident. The first time my wife drove the '40 after the accident she was very nervous, but after a few more trips she was back up to her fast driving again. (Everyone knows she has a "heavy foot"!)

The accident was a harrowing incident and did lots of damage to many vehicles, but we are very glad no one was badly injured.

Triple Axle Debacle
By Wings Kalahan
Baldwin, WI

As of 2019, I have been the announcer for the National Street Rod Association (NSRA) for 42 years. I call my show "Cruisin'with Wings." I travel around the United States to the ten NSRA shows in a forty-foot 1988 Roadway trailer outfitted with a studio, broadcasting equipment and sleeping quarters. Traveling these thousands of miles, it's a wonder I haven't had many accidents and mechanical problems.

In Oregon in 1988 or '89, I was driving my modified '38 Chevy ton and a half truck (since retired from duty) which was pulling my trailer when the planetary gears in the automatic transmission "blew." Then while crossing the Arizona desert the 70-gallon main fuel tank cracked and spewed out several gallons of fuel. I left the truck in Flagstaff, pulled off the tank and took it to my friend's rod shop in Phoenix where we attempted to repair it. He put dry ice in the tank and welded it. I was very skeptical, to say the least, about this technique but it turned out to be an excellent repair.

In Prescott, Arizona a front tire exploded! I limped on the rim into a truck stop where I priced a used tire and wheel at $275, but managed to rent it for $40, so was good to go!

I also had a crack in a tubing exhaust header that in turn burned the starter severely. I later went back to cast iron manifolds and picked up a "big block" starter along the way.

The several incidents mentioned above were bad enough, but in 2010 I had my most serious incident while traveling from the Street Rod Nationals North in Kalamazoo, Michigan to the Northeast Street Rod Nationals in Burlington, Vermont. I was parked in a rest stop near Rochester, New York and had bedded down for the night. A Freightliner "straight truck" was parked perpendicular to my rig 150' away on slightly higher ground. In retrospect, I realized that the truck's e-brake air button was not set properly which apparently was the major reason this rig "got loose" and rolled down and collided with my trailer. The trailer, which weighs 18,000 pounds was moved about five feet. The incident destroyed the center axle of the three-axle setup of the trailer and did considerable damage to the other two as well.

The truck driver, who was asleep in his sleeper cab, of course woke up and was pretty upset about it all, what with the harm to my trailer and to his radiator, which was severely damaged. Luckily, I had a spare axle with me so when everybody calmed down and the tow truck arrived, I asked the driver, who was familiar with the area, where I could have the center axle "changed out." This was done and the other two axles were "cold straightened." All in all, with the axle and other body damage, there was $16,000 worth of damage done to the trailer! After the repair I was able to continue the trip to Vermont to do that show and then on home to Wisconsin. Another show in Sacramento California followed that one. On this latter trip one of the end axles broke and had to be welded in a local shop. I later changed all three axles to help prevent further problems.

This covers some of the main incidents over my 30 years of traveling. I will be logging lots of miles for a few years to come and am looking forward to accident free traveling. It's always good to have a trip free of surprises!

ACKNOWLEDGEMENTS

Thank you to the many who contributed to my book.

The 51 street rodders who gave of their time to relate their stories to me.
Editor: Melanie Stultz-Backus
Beta readers:
    Don Lefebvre
    Todd Cromie
    Kim Cromie
    Wanda Baillargeon
    Marion Stultz
Cartoon artists:
    Mariel Stultz Klingbeil
    Matthew FitzGerald